Smart Biosensors in Medical Care

Advances in Ubiquitous Sensing
Applications for Healthcare

Smart Biosensors in
Medical Care

Volume **8**

Edited By

Jyotismita Chaki

School of Information Technology and Engineering,
Vellore Institute of Technology, Vellore, India

Nilanjan Dey

Department of Information Technology,
Techno India College of Technology, Kolkata, India

Debashis De

Department of Computer Science and Engineering,
West Bengal University of Technology, Kolkata, India

Series Editors

Nilanjan Dey
Amira S. Ashour
Simon James Fong

ELSEVIER

ACADEMIC PRESS
An imprint of Elsevier

Academic Press is an imprint of Elsevier
125 London Wall, London EC2Y 5AS, United Kingdom
525 B Street, Suite 1650, San Diego, CA 92101, United States
50 Hampshire Street, 5th Floor, Cambridge, MA 02139, United States
The Boulevard, Langford Lane, Kidlington, Oxford OX5 1GB, United Kingdom

Library of Congress Cataloging-in-Publication Data
A catalog record for this book is available from the Library of Congress

British Library Cataloguing-in-Publication Data
A catalog record for this book is available from the British Library

ISBN: 978-0-12-820781-9

For information on all Academic Press publications
visit our website at https://www.elsevier.com/books-and-journals

Publisher: Mara Conner
Acquisitions Editor: Fiona Geraghty
Editorial Project Manager: John Leonard
Production Project Manager: Kamesh Ramajogi
Designer: Christian J. Bilbow

Typeset by Thomson Digital

Working together
to grow libraries in
developing countries

www.elsevier.com • www.bookaid.org

Contents

Contributors

Geetika Aggarwal
Northumbria University, Newcastle upon Tyne, United Kingdom

Alaknanda Ashok
Graphic Era Deemed to be University, Dehradun;
G.B. Pant University of Agriculture and Technology, Pantnagar, India

Maanvi Bhatnagar
Department of Electrical & Electronics Engineering, Birla Institute of
Technology, Mesra, Ranchi, Jharkhand, India

Richard Binns
Northumbria University, Newcastle upon Tyne, United Kingdom

Xuewu Dai
Northumbria University, Newcastle upon Tyne, United Kingdom

B. Sri Sai Deepthi
Mamatha Medical College, Khammam, India

Pradeep Kumar Dewangan
Department of Chemistry, National Institute of Technology, Raipur, India

Ankur Dumka
Graphic Era Deemed to be University, Dehradun;
G.B. Pant University of Agriculture and Technology, Pantnagar, India

Bharat Gupta
National Institute of Technology, Patna, India

Gauri Shanker Gupta
Department of Electrical & Electronics Engineering, Birla Institute of
Technology, Mesra, Ranchi, Jharkhand, India

Dharm Singh Jat
Namibia University of Science and Technology, Windhoek, Namibia

Vijayalakshmi Kakulapati
Sreenidhi Institute of Science and Technology, Hyderabad, India

Faruk Kazi
Veermata Jijabai Technological Institute, Mumbai, India

Fahmida Khan
Department of Chemistry, National Institute of Technology, Raipur, India

Sudhansu Kumar Mishra
Birla Institute of Technology, Mesra, Ranchi, Jharkhand, India

Sweta Kumari
Birla Institute of Technology, Mesra, Ranchi, Jharkhand, India

Anton S. Limbo
Namibia University of Science and Technology, Windhoek, Namibia

Sheri Mahender Reddy
Sreenidhi Institute of Science and Technology, Hyderabad, India

Namrata Misra
School of Biotechnology; KIIT Technology Business Incubator, Kalinga Institute of Industrial Technology (KIIT), Deemed to be University, Bhubaneswar, India

Dusmanta Kuamar Mohanta
Department of Electrical & Electronics Engineering, Birla Institute of Technology, Mesra, Ranchi, Jharkhand, India

Maheswata Moharana
Hydro & Electrometallurgy Department, CSIR-Institute of Minerals and Materials Technology, Bhubaneswar, India

Subrat Kumar Pattanayak
Department of Chemistry, National Institute of Technology, Raipur, India

Sahar Qazi
Department of Computer Science, Jamia Millia Islamia, New Delhi, India

Mayur Rathi
Walchand College of Engineering, Sangli, India

Khalid Raza
Department of Computer Science, Jamia Millia Islamia, New Delhi, India

Reza Saatchi
Sheffield Hallam University, Sheffield, United Kingdom

Susrita Sahoo
School of Biotechnology, Kalinga Institute of Industrial Technology (KIIT), Deemed to be University, Bhubaneswar, India

Satya Narayan Sahu
School of Biotechnology, Kalinga Institute of Industrial Technology (KIIT), Deemed to be University, Bhubaneswar, India

Sitanshu Sekhar Sahu
Birla Institute of Technology, Mesra, Ranchi, Jharkhand, India

Charu Singh
Sat-Com (PTY) Ltd., Windhoek, Namibia

Rakesh Kumar Sinha
Department of Bio-Engineering, Birla Institute of Technology, Mesra, Ranchi, Jharkhand, India

Shefali Sonavane
Walchand College of Engineering, Sangli, India

Mrutyunjay Suar
School of Biotechnology; KIIT Technology Business Incubator, Kalinga Institute of Industrial Technology (KIIT), Deemed to be University, Bhubaneswar, India

Sunil Tamhankar
Walchand College of Engineering, Sangli, India

João Manuel R.S. Tavares
Instituto de Ciência e Inovação em Engenharia Mecânica e Engenharia Industrial, Departamento de Engenharia Mecânica, Faculdade de Engenharia, Universidade do Porto, Porto, Portugal

Padma Rani Verma
Department of Chemistry, National Institute of Technology, Raipur, India

About the editors

Nilanjan Dey is an Assistant Professor (Senior Grade) in the Department of Information Technology at Techno India College of Technology (under Techno India Group), Kolkata, India. He completed his PhD in 2015 from Jadavpur Univeristy, Kolkata, India. He is a Visiting Fellow of Wearable Computing Laboratory, Department of Biomedical Engineering, University of Reading, UK. He is the Visiting Professor of the College of Information and Engineering, Wenzhou Medical University, People's Republic of China and Duy Tan University, Vietnam. He has held the honorary position of Visiting Scientist at Global Biomedical Technologies Inc., California, USA (2012–2015). He is a Research Scientist at the Laboratory of  Applied Mathematical Modeling in Human Physiology, Territorial Organization of Scientific and Engineering Unions, Bulgaria; Associate Researcher of Laboratoire RIADI, University of Manouba, Tunisia; and Scientific Member of Politécnica of Porto. Before joining Techno India College of Technology, he served as an Assistant Professor at JIS College of Engineering and Bengal College of Engineering and Technology.

With more than 10 years of teaching and research experience, he has authored/edited more than 40 books with Elsevier, John Wiley & Sons, CRC Press, and Springer, and has published more than 350 research articles. His h-index is 30 with more than 4000 citations. He is the Editor-in-Chief of the *International Journal of Ambient Computing and Intelligence* (IJACI, IGI Global, UK, Scopus) and the *International Journal of Rough Sets and Data Analysis* (IGI Global, US, DBLP, ACM dl). He is the Series Co-Editor of *Springer Tracts in Nature-Inspired Computing* (STNIC) and *Advances in Ubiquitous Sensing Applications for Healthcare* (AUSAH), Elsevier; Series Editor of *Computational Intelligence in Engineering Problem Solving and Intelligent Signal Processing and Data Analysis*, CRC Press (FOCUS/Brief Series), De Gruyter Series on the Internet of Things and Advances in Geospatial Technologies (AGT) Book Series (IGI Global), United States; serves as an editorial board member of several international journals, including the *International Journal of Image Mining* (IJIM), Inderscience; Associate Editor of *IEEE Access* (SCI-Indexed); and the *International Journal of Information Technology*, Springer.

In addition, he was recognized as one of the top 10 most published academics in the field of Computer Science in India during the period of consideration, 2015–17, at the Faculty Research Awards organized by Careers 360 at New Delhi, India, on March 20, 2018.

His main research interests include medical imaging, machine learning, and computer aided diagnosis as well as data mining. He has been on the program committees of over 50 international conferences, has acted as a program co-chair and/or advisory chair of more than 10 of them, and as an organizer of five workshops.

He has given more than 50 invited lectures in 10 countries, including many invited plenary/keynote talks at international conferences such as ITITS2017 (China), TIMEC2017 (Egypt), and BioCom2018 (UK).

Jyotismita Chaki is an Assistant Professor in the School of Information Technology and Engineering at Vellore Institute of Technology, Vellore, India. She did her PhD (Engineering) from Jadavpur University, Kolkata, India. Her research interests include computer vision and image processing, pattern recognition, medical imaging, soft computing, data mining, and machine learning. She has authored many international conferences and journal papers. She is also the author of the following books: *A Beginner's Guide to Image Preprocessing Techniques* (CRC Press, Taylor and Francis) and *A Beginner's Guide to Image Shape Feature Extraction Techniques* (CRC Press, Taylor and Francis). She has served as a reviewer of *Applied Soft Computing* (Elsevier), *Biosystem Engineering* (Elsevier), *Pattern Recognition Letters* (Elsevier), *Journal of Visual Communication and Image Representation* (Elsevier), *Signal Image and Video Processing* journal (Springer), and also served as Program Committee member of the 2nd International Conference on Advanced Computing and Intelligent Engineering 2017 (ICACIE-2017) and 4th International Conference on Image Information Processing (ICIIP-2017).

Debashis De earned his M.Tech from the University of Calcutta in 2002 and his PhD (Engineering) from Jadavpur University in 2005. He is the Professor and Director of the Department of Computer Science and Engineering at the West Bengal University of Technology, India, and Adjunct Research Fellow at the University of Western Australia. He is a senior member of the IEEE, Life Member of CSI, and a member of the International Union of Radio Science. He worked as an R&D engineer for Telektronics and a programmer at Cognizant Technology Solutions. He was awarded the prestigious Boyscast Fellowship by the Department of Science and Technology, Government of India, to work at Herriot-Watt University, Scotland, UK. He received the Endeavour Fellowship Award during 2008–2009 by DEST Australia to work at the University of Western Australia. He also received the Young Scientist award in 2005 at New Delhi and again in 2011 at Istanbul, Turkey, from the International Union of Radio Science, Brussels, Belgium. His research interests include mobile cloud computing and green mobile networks. He has published in more than 250 peer-reviewed international journals from IEEE, IET, Elsevier, Springer, World Scientific, Wiley, IETE, Taylor Francis, and ASP; 70 International conference papers; four research monographs from Springer, CRC, and NOVA; and 10 textbooks published by Pearson education. He is Associate Editor of the journal *IEEE ACCESS* and an Editor of *Hybrid Computational Intelligence*.

Preface

Due to countless applications in clinical biochemistry and even in environmental and science analytical chemistry, biosensors are in very high demand. Because of the quantity of applications in major fields, this recent development has gained a lot of interest. Biosensors are devices that transform biological information into signals that are analytically useful. Biosensors offer improved specificity and high sensitivity signals that are easy to detect. Today, with the support of biosensors, you can know what's happening inside your body just by sitting at home. Biosensors have substituted for the slow, old, and painful testing procedures and offer a new and responsive mechanism that is easy to use. Such testing can boost the efficiency of patient care. As an example, there is a blood-glucose sensor which is broadly used in the medical field for the diabetic patient. This book provides comprehensive information of up-to-date requirements in hardware, communication, and calculating for next-generation medical care systems. Detailed information on various recent worldwide system operations is presented. In particular, challenges using sensors in medical care are highlighted. The purpose of this book is to not only help beginners with a holistic approach toward understanding medical care systems but also to present to researchers information about new technological trends and design challenges they have to cope with while designing such systems. This book is concerned with supporting and enhancing the utilization of smart biosensors in medical care. It provides a knowledgeable forum to discuss the characteristics of biosensors in the different domains of medical care. This book is proposed for professionals, scientists, and engineers who are involved in the new techniques of medical care systems. It provides an outstanding foundation for undergraduate and post-graduate students as well. It has several features, including (1) an outstanding basis of biosensor-based medical care data analysis and management, (2) different data transmission techniques and applications, with extensive studies for the biosensor-based medical healthcare system, and (3) various challenges of biosensor based medical healthcare system.

The book is organized as follows.

Chapter 1 deals with the comparative analysis of various feature extraction techniques and classification algorithms that are best suited for obtaining motor imagery signals. This chapter discusses linear features such as power spectral density and band power, as well as mean and maximum power of motor imagery signals, along with some nonlinear features such as correlation coefficient and approximate entropy. The different linear and nonlinear features from EEG signals are also discussed.

Chapter 2 deals with the importance of biosensors in diagnosing several infectious as well as pathogenesis diseases like malaria. The identification of disease markers and predicting their effects by computational approaches has the potential to generate personalized tools for the diagnosis and treatment of malaria. Previous malarial studies have stated that both PfDHFR-TS and PfHDP are promising biomarker targets. Against this backdrop, an in-depth analysis of the molecular interaction of

PfHDP and PfDHFR with HEME is studied by using a computational approach. The current scenario of the biosensor as discussed in this chapter along with the computational study provide useful insights for the design of therapeutic interventions for infectious diseases.

Chapter 3 focuses on the development and research of various types of biomarkers, which are basically used for the detection and diagnosis of diseases. The interaction of different proteins (PDB ID: 4MPF and 3D78) with halauxifen-methyl, mesosulfuron-methyl andibutyl benzene-1,2-dicarboxylate, and 1,2-dimethoxy-4-prop-2-enylbenzene by the molecular docking method is examined. The impact of nanotechnology on biosensors is discussed. Also, the influence of optical- and electrochemical-based biosensors in medical care are presented.

The focus in Chapter 4 is an insight into all the smart biosensors which are actively being used in maintaining a point of care (PoC) lifestyle. Different types of bio-receptors in biosensors like nucleic acids/enzymes, aptamers, antibodies, ions, molecules, nanoparticles, organisms/cells, tissues, and so on, and different kinds of biological transducers like electrochemical biotransducers, ion channel switch biotransducers, reagentless fluorescent biotransducers, and so forth, are discussed in this chapter. Also, various types of electrochemical biosensors like amperometric biosensors, potentiometric biosensors, impedimetric biosensors, and voltammetric biosensors; and different types of optical biosensors, electrical biosensors, piezoelectric biosensors, and thermometric biosensors are presented. This chapter provides an overview of different applications of biosensors.

Chapter 5 reviews methods of energy harvesting techniques through human activities and the use of that technology in biological applications (healthcare applications) as well as consumer electronics. This chapter provides an overview with a brief comparison of various available energy harvesting technologies and their power consumption in extending the battery lifetime. The wireless body area network and its application in medical healthcare is also discussed.

Chapter 6 experimentally demonstrates the BER (bit error rate) performance comparison of a novel electroencephalography (EEG) healthcare system using an 8-pixel and 16-pixel OLED screen and DSLR camera as transmitter and receiver, respectively. The reasons for using VLC; the flaws in radio frequency technology in healthcare, in addition to frequency crunch, like safety for human health, security, and low-cost implementation; and green wireless communication technology are discussed. This chapter also gives an overview of different types of receivers in the VLC system like the photodetector and camera. The application of VLC in healthcare is included.

Chapter 7 focuses on active queue management (AQM) based schemes using fuzzy logic in order to provide high utilization of network with low losses and delay within the network—thus providing effective control in the human body area network (HBAN) system and providing smart e-health applications using biosensors. Different protocols of AQM methodology like trail drop, random early detection, random exponential marking (REM), blue queue, REM controller, and adaptive virtual queue are discussed in this chapter. Also, the application of fuzzy-logic-based congestion control protocol in the HBAN system for smart e-health using biosensors is presented.

Chapter 8 provides an overview of biofabrication of graphene quantum dots as a fluorescent nanosensor for detection of toxic and heavy metals in biological and environmental samples. The authors reported the one-step facile microwave synthesis of blue bright fluorescent N-GQDs by using glucose and urea as source materials for the detection of heavy metal ions. This method does not require any expensive reagents, thus making it an environmentally friendly and cost-effective approach.

Chapter 9 provides an overview of machine learning analysis of topic modeling reranking of clinical records. In this chapter the authors used a method that combines Statistical Topic Models, Language Models, and Natural Language Processing, in order to retrieve clinical records. The decomposition of clinical record summaries into topics which enable the effective clustering of relevant documents, based on the topic under study, is explored.

Chapter 10 compares various biosensors for the smart sensing of medical disorders for the physically impaired person. It focuses on how those sensors can be used along with wearable body sensors within a system. In any sensor network, the communication link is a vital medium to transfer data with transport protocol. Different attributes to measure the network performance in maximizing the accuracy of the system like transmission efficiency, network active time, data consistency, latency, visualization, multitenancy, and quick response are discussed. Also brought up are the data pre-processing techniques and the techniques for storing data in the cloud for secure access. In addition, an application for a blind person with wearable sensors and a smart stick is presented.

The focus in Chapter 11 is on a speech-based automation system to assist patients in an orthopedic ward to automate processes using voice commands. How to create the automated system is demonstrated using two Raspberry Pis devices, where one device represents a computing node that will be placed in a ward and the second device represents the monitoring device to be placed at the nurses' station. A speech recognition system employs two main types of algorithms to achieve speech recognition: Hidden Markov Model (HMM) and Artificial Neural Networks (ANN) and these are discussed in this chapter.

Editors
Dr. Nilanjan Dey
Techno India College of Technology, India

Dr. Jyotismita Chaki
Vellore Institute of Technology, India

Dr. Debashis De
West Bengal University of Technology, India

Prototype algorithm for three-class motor imagery data classification: a step toward development of human–computer interaction-based neuro-aid

Gauri Shanker Gupta[a], Maanvi Bhatnagar[a], Dusmanta Kuamar Mohanta[a], Rakesh Kumar Sinha[b]

[a]Department of Electrical & Electronics Engineering, Birla Institute of Technology, Mesra, Ranchi, Jharkhand, India; [b]Department of Bio-Engineering, Birla Institute of Technology, Mesra, Ranchi, Jharkhand, India

1 Introduction

Medical systems have been striving to provide better quality of life to humanity at large. Particularly with the advent of brain–computer interface (BCI), there has been a paradigm shift to use different brain signals for assisting physically challenged people. Motor imagery (MI)-based BCI systems provide significant rehabilitation for people suffering from spinal cord injuries, amyotrophic lateral sclerosis (ALS) [1–3]. Electroencephalogram (EEG) signals have emerged to be the most suitable for BCI systems. Researchers have concluded that the same area of motor cortex is responsible for both MI and actual motor execution (ME) [4,5]. Hence, out of all the existing techniques available for BCI operation, sensorimotor-based recognition systems obtained using EEG have gained popularity. These types of systems work based on user's intent. Whenever a person performs or imagines a limb movement, there is a change in EEG activity observable in specific brain rhythms (alpha and beta waves) due to changes in the firing pattern of underlying neurons. The decrease or increase of action potentials due to deactivation and activation of the neuronal population is known as event-related desynchronization (ERD) and event-related synchronization (ERS), respectively [6]. These phenomena are highly frequency selective and MI signals capture such phenomena which need to be extracted suitably for pragmatic purposes.

Efficient recognition of MI signals can prove to be a breakthrough in the medical field for successful rehabilitation of physically challenged people. There have been research efforts toward finding effective ways to identify these signals for real-time applications. However, the accuracy of classification techniques is still not up to the mark. In addition, computational time for classification is also on a higher side for real-time applications. These two factors need to be addressed especially for patients who are undergoing continuous rehabilitation sessions using BCI. Hence, algorithms providing good accuracy are very essential.

Several approaches have been proposed in the literature to improve the classification accuracy having an acceptable computational time. For instance, Yong and Menon [7] proposed classification of binary-class and three-class rest, imagery grasp, and elbow movement signals recorded from the same limb. Filter Bank Common Spatial Pattern (CSP) and logarithmic band power (BP) features were extracted. Linear discriminant analysis (LDA), logistic regression (LR) using a fast Bayesian method, and support vector machine (SVM) with radial basis function (RBF) kernel was used to classify the three-class and binary-class data using the tenfold cross-validation scheme. However, this methodology leads to a very low classification of about 66.9%. The main problem lies in the selection of an appropriate kernel function for SVM classifier because of the nonstationary nature of EEG signals. To get rid of this problem Atkinson and Campos [8] used SVM with particle swarm optimization (PSO) technique to optimize the parameters of the kernel function for classification of emotion recognition using EEG signals. Murugappan et al. analyzed human emotions data of happiness, sadness, disgust, etc., using wavelet features and k-nearest neighbors (k-NN) and LDA classifiers and acquired an average accuracy 83.26% [9]. Yalcin et al. studied detection of epilepsy using EEG signals using artificial neural network and optimized its parameters using PSO and achieved a classification accuracy of around 87.23% [10]. Zhang et al. conducted studies on seizure signals obtained from EEG recordings and obtained a classification accuracy of 98.10% using SVM, LDA, and k-NN classifier and local mean decomposition features [11]. Further, efficacy of a BCI system highly depends on the type of preprocessing and feature extraction algorithms used [12,13]. McFarland et al. and Koles et al. proposed spatial filtering techniques such as spline surface Laplacian (SSL), small Laplacian, and common average reference (CAR) for filtering of EEG signals recorded during control of cursor on a computer [14,15]. The spatial filters were found efficient for artifact removal. Townsend et al. and Blankertz et al. utilized the time–frequency-based features such as BP [16] and power spectral density (PSD) [17] to quantify MI signals. Corsi et al. tried a new approach by mixing EEG and MEG signals together and extracting power features from them. They further used a hybrid classifier involving LDA and Bayesian techniques to improve the accuracy [18]. Different feature extraction, preprocessing, and classification techniques discussed here are implemented on three-class MI framework in this research work.

The proposed research work attempts to enhance accuracy and reduce computational time for BCI system along with analyzing the linear as well as nonlinear behavior of these signals for a multi-class framework [19,20]. The multi-class framework has inherent advantage of increasing the degrees of freedom for neuro-aid device through pragmatic applications of BCIs. Thus, the other innovative contribution of this chapter is the multi-class framework pertaining to MI for right hand (RH), left hand (LH), and foot along with their analysis of linear and nonlinear behavior. Special emphasis has been given to all these stages to get a robust BCI framework [21,22].

2   Methods

Proposed methodology used in this work is as given in Fig. 1.1. This framework has three important processing steps, namely filtering, feature extraction, and classification. The raw EEG signals are firstly filtered and then the various feature extraction techniques are applied to extract important features from the data. These features are then fed to a classifier. The following subsections contain all these steps with detailed explanation.

2.1   Signal acquisition and preprocessing

2.1.1   Signal acquisition

The first step in this study is to obtain the EEG data. For this purpose, dataset IIIa from the BCI Competition III conducted by University of Graz is used in this study [22]. It consists of recorded data from four classes, namely LH, RH, foot, and tongue MI signals obtained from three subjects, namely K3b, K6b, and L1b having dissimilar experiences in imagery recording. Data were recorded with a sampling rate of 250 samples per second using a 64 channel Neuroscan amplifier of which only 60 electrodes were used for recording. Fig. 1.2 shows a prototype of the EEG recording

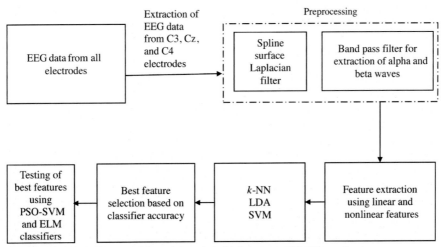

FIGURE 1.1 Block diagram of the proposed methodology used in this work.

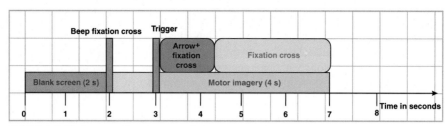

FIGURE 1.2 Timing paradigm for recording scheme.

cap. The details of the recording paradigm are as given [23]. Subject K3b consists of recording over 45 trials and K6b and L1b have recorded data for 30 trials. The data recordings lie in the frequency range of 1–50 Hz. In this work, only C3, Cz, and C4 electrodes are used for further analysis because these electrode positions correspond to those areas of motor cortex that are responsible for generation of MI signals [24]. Fig. 1.3 demonstrates the EEG recording of all the C3 electrodes for one trial (1000 samples) of K3b subject. Also, only data from three classes are used, that is, LH, RH, and foot because tongue data are highly nonstationary and nonlinear in nature.

2.1.2 Preprocessing

Preprocessing is one of the important stages in BCI systems. Raw EEG signals consist of information carrying signals along with some unwanted signals such as noise and artifacts arising from various sources such as muscle movement and eye blink [25]. Therefore, to improve the spatiotemporal resolution of the signal and remove unwanted noise components, spatial filtering using SSL algorithm which allows the calculation of surface potential of the signals using the spherical coordinate system has been applied [26]. It is calculated with the help of SSL toolbox [27] as shown

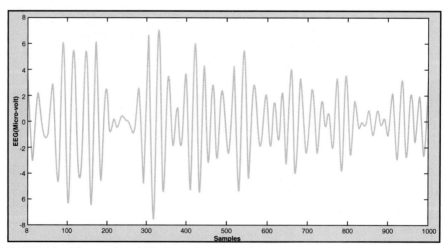

FIGURE 1.3 EEG recording of C3 electrode for one trial.

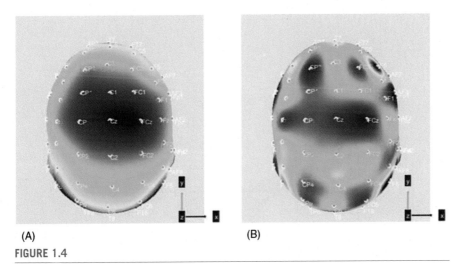

(A)   (B)

FIGURE 1.4

(A) Scalp plot for unfiltered EEG data. The EEG potential is distributed only over a particualr area as shown by the *blue* colour, and (B) scalp plot of filtered EEG signal using SSL filtering technique with an even distribution of EEG potential over the entire scalp.

in Fig. 1.4A and B, respectively, for unfiltered and filtered EEG signals. According to studies conducted in [26], it is concluded that Laplacian filters have good SNR in comparison with other techniques such as CAR. As ERD/ERS-dependent MI phenomenon is highly frequency-selective selection of correct frequency band is also important [12]. For this purpose, a Butterworth filter of order 6 has been used to obtain both the alpha band (8–13 Hz) and beta band (14–30 Hz). The upper and lower cutoff frequencies of band pass filter for obtaining alpha band are 7.5 Hz and 13.5 Hz, respectively. For beta band the upper and lower cutoff frequencies are 14.5 Hz and 30.5 Hz, respectively. The stopband attenuation and passband ripple are 40 dB and 2.8 dB, respectively. The raw EEG and filtered alpha and beta band signals are shown in Fig. 1.5A–C, respectively. A low-order Butterworth filter has been chosen so as to get a stable filter having less computational complexity and minimum error due to quantization noise. According to studies, a filter with order 4–8 works best in combination with all the feature extraction techniques [28]. A good preprocessing technique can improve the performance of a good BCI while using the same feature extraction and classification algorithms.

2.2   Feature extraction algorithms

The data used are a time series having scattered power components. Therefore, features based on obtaining the power of the signals as a function of time or frequencies are best suited in this case [16,17]. Linear features such as the BP features, PSD features, and max and mean power features are used for analysis purpose along with nonlinear features such as the approximate entropy and correlation coefficient. As

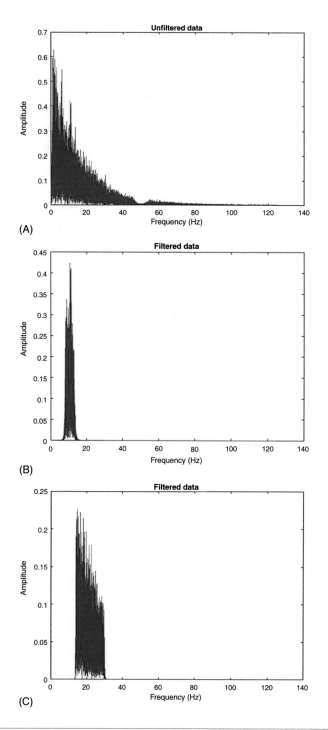

FIGURE 1.5

(A) Unfiltered raw EEG signal of subject K3b corresponding to C3, C4, and Cz electrode,
(B) filtered EEG signal, alpha band (8–13 Hz), and (C) filtered EEG signal, beta band (14–30 Hz).

both alpha and beta bands are taken into consideration, therefore, feature vectors obtained from both the bands are merged together. Let f be the feature space then the feature matrix for the three electrodes {C3, C4, Cz} for a class and all the trials is given as:

$$f = \left\{ f_1^{e1}, f_2^{e1}, f_3^{e1} \cdots f_n^{e1}, f_1^{e2}, f_2^{e2}, f_3^{e2}, \cdots f_n^{e2}, f_1^{e3}, f_2^{e3}, f_3^{e3} \cdots f_n^{e3} \right\}^T \qquad (1.1)$$

n represents the number of trials and $e1$, $e2$, $e3$ represents the three electrodes. From the feature space, features from any two electrodes are selected at a time for classification purpose. As discussed before, C3 electrode corresponds to an occurrence of ERD upon LH MI; similarly, C4 and Cz electrodes correspond to occurrence of ERD upon LH and foot MI, respectively. Therefore, the feature vectors obtained correspond to that of LH versus RH, LH versus foot, and RH versus foot. This clearly indicates that a set of three-feature vector is obtained for each subject. There are 45 trials each for training data and test data per class for K3b subject and 30 trials each for training and test data per class for K6b and L1b subjects. One feature per trial is calculated for each electrode for each class for every subject. The sampling rate is 250 sample/s and the MI dataset is 4 s long. The number of samples per trial is 1000. Features are calculated by averaging over the 1000 samples of each trial and hence one feature per trial is obtained for each electrode. Therefore, K3b subject has three 45 × 2 feature vectors per class for training data and test data each. Subjects K6b and L1b have three 30 × 2 feature vectors for training data and test data each. The programming was done on MATLAB platform.

2.2.1 Linear features

2.2.1.1 Band power features

The advantage of using this technique is that computationally it is the most efficient technique for distinguishing between signals obtained from different imagery conditions. The BP is calculated using Eq. (1.2) for signal x_n, which represents the data recordings of an electrode over N ($N = 1000$) samples.

$$P_{B.P} = \frac{1}{N} \sum_{n=1}^{N} x_n^2 \qquad (1.2)$$

This process is repeated for every trial, that is, 45 times for K3b subject and 30 times for K6b and L1b subject for all the three electrode combinations. Three feature vectors (LH vs. RH, LH vs. foot, and RH vs. foot) are obtained corresponding to each subject for training data and test data in accordance with vector as represented in Eq. (1.1) for all the subjects.

2.2.1.2 Power spectral density features

PSD of a signal gives an analysis of the distribution of power over the entire frequency range. The main objective of using this method is to obtain the spectral density estimation from the given data. It is estimated by calculating the Fourier transform (FT) of the signals' autocorrelation function. It perceives the signal as a stochastic process and then determines its power. In the present work, both parametric and nonparametric approaches are used. In parametric approach, it is assumed that modeling of the signal is done by considering it stationary in that time window having minimum spectral leakage. Burg and Yule–Walker auto-regressive (AR) parametric models are used [29]. A sixth-order AR model is used for obtaining the model coefficients [29]. The corresponding PSD is obtained using Eq. (1.3) for both the techniques.

$$P_{xx}(nT) = \frac{E_p}{\left|1 + \sum_{n=1}^{p} a_p(k) e^{-i2\pi fnT}\right|^2} \tag{1.3}$$

Here, P_{xx} is the estimated PSD, E_p is the mean square error coefficient, a_p are the model parameters, f represents the sampling rate, and p is the order of the model. Burg PSD focuses on minimization of forward and backward prediction errors while keeping in mind the Levinson–Durbin recursion. It does not require the calculation of autocorrelation function. Whereas in Yule–Walker method a biased estimate of the signal's autocorrelation function is obtained. The least squares of the forward prediction errors are minimized for achieving this.

The nonparametric methods used are periodogram and Welch. In the periodogram method a small-time window from $t = 3$ s to $t = 7$ s is considered. FT of the biased estimate of the autocorrelation sequence of this window is obtained using Eq. (1.4) and averaged over all samples of a trial. Here, a rectangular window is chosen to obtain a one-sided periodogram.

$$P(nT) = \frac{\Delta t}{N} \left|\sum_{n=0}^{N-1} x_n e^{-i2\pi fnT}\right|^2 \tag{1.4}$$

Here, Δt is the sampling interval, and $N(N = 1000)$ is the number of samples of the signal x_n which represents the data recordings for any of the three electrodes for one trial, sampled at a frequency f_s.

Welch's technique is an improved version of the periodogram method. Signal is divided into equal segments and to each segment hamming window is applied to reduce the variance of the periodogram. Average over all the trials for each windowed segment is taken and its periodogram is calculated to give an estimate of the PSD [30]. The PSD is calculated with the help of Eqs. (1.5) and (1.6). With the help of Eq. (1.5) periodogram of a single trial is calculated.

$$P_i(nT) = \frac{1}{NU} \left|\sum_{n=0}^{N-1} w[n] x_i[n] e^{-2j\pi fnT}\right|^2 \tag{1.5}$$

Here, U is the normalization constant, $w[n]$ is the windowing function (hamming window), and $x_i[n]$ is the input signal obtained from data recorded from any one of the electrodes of the ith trial.

Features of all the trials are obtained similarly and concatenated into a feature vector in accordance with the feature vector as represented in Eq. (1.1). Three features vectors (LH vs. RH, LH vs. foot, and RH vs. foot) are obtained for each subject with a size 45 × 2 for subject K3b and 30 × 2 for subjects K6b and L1b for both training and testing data.

2.2.1.3 Features based on maximum power and mean power

Maximum power is obtained by taking into consideration the maximum amplitude of the signal $x(n)$. Signal $x(n)$ represents the data recordings from any of the three electrodes. It is obtained as per the equation: Max $P = \max(x^2(n))$. Similarly, mean power is computed by averaging the overall signal power over all the samples of a trial, that is, mean $P = \text{mean }(x^2(n))$. The features hence obtained are arranged as a vector in accordance with Eq. (1.1).

2.2.3 Nonlinear features

2.2.3.1 Approximate entropy

Entropy, as a concept that a value would be reasonably considered from a series in an ordered system, can be defined as kind of index of symmetry or the degree of arbitrariness. The entropy will have a complex value if the number of categorizations in a series is more intricate or without any order, and vice versa. These features are concatenated according to Eq. (1.1).

2.2.3.2 Correlation coefficient

The coefficient of correlation is a numerical measure of some kind of correlation, denoting a statistical association amongst two given variables. These variables are the two columns of a given dataset of observations, called a sample, or two mechanisms of a multivariate random variable with a known distribution.

All the features obtained from the aforementioned techniques are then fed to the classifiers for classification purpose. Features obtained using parametric and nonparametric linear techniques along with nonlinear techniques are shown in Fig. 1.6A–C, for all the trials of electrode C3 for subject K3b.

2.2.3.3 Statistical validation of extracted features

The statistical analysis is only restricted to testing of linear features as the comparison has to be made between the best linear features with any nonlinear features.

- *t-test*: Paired sample *t*-test is used to authenticate whether there is any significant difference between the signals obtained from electrode 1 and electrode 2 while considering the same MI signals over both the electrodes. This test is performed in order to validate the phenomenon of ERD/ERS occurring simultaneously on both the electrodes that are located on opposite sides of the brain. The *t*-test can be implemented under normality assumption and is frequently used in BCI studies [29,30]. A null hypothesis testing was done at a significance level of $P = 0.05$ and the null hypothesis is that the data obtained from both the electrodes come from the same normal distribution with zero mean and some unknown variance. For instance, BP features were considered, and the

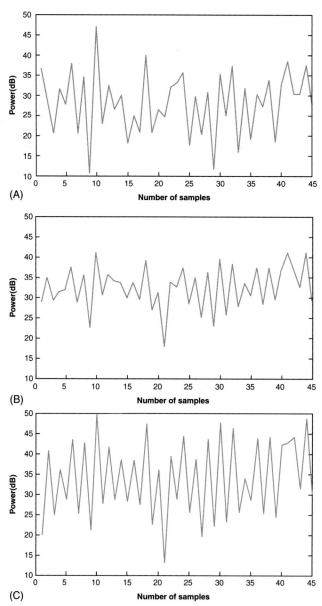

FIGURE 1.6 Plots showing features obtained using different feature extraction techniques for subject K3b for electrode C3.

(A) Burg PSD, (B) correlation coefficient, and (C) approximate entropy.

t-test was conducted between the features vectors obtained while considering the LH (C4) versus RH (C3) electrode pair combination. The difference or similarity between the data obtained from the electrodes was evaluated using the hypothesis test as mentioned before at a significance level of $P = 0.05$ and the value of h is obtained, which is the hypothesis test result. The value of $h = 1$ result in rejection of null hypothesis and $h = 0$ indicates its acceptance. This test was conducted on all feature vectors obtained using the feature extraction techniques discussed earlier for all the subjects and all electrode pair combinations. This test helps in evaluation of the significance of the various algorithms used for both feature extraction purpose and classification purpose.

- *Friedman test*: This test is a nonparametric test of two-way analysis of variance. It is applied to obtain the best features which can be given as an input to the classifier in order to obtain good classification accuracy [30]. This test was performed by considering two features at a time, for example, BP features and Burg PSD features, with multiple observations of each feature. Both the feature vectors are concatenated columnwise. Then it is tested that do the two features under comparison show the same nature and have similar properties with the help of *p*-value which is defined as the probability of accepting the null hypothesis as true or false. Different p and x^2 values are obtained for different feature pairs based on which the hypothesis is rejected or accepted. This test was conducted for the following feature pairs:
  - BP versus Yule–Walker PSD/periodogram PSD/Welch PSD/Burg PSD/max power/mean power.
  - Yule–Walker PSD versus periodogram PSD/Welch PSD/Burg PSD/max power/mean power.
  - Periodogram PSD versus Welch PSD/Burg PSD/max power/mean power.
  - Welch PSD versus Burg PSD/max power/mean power.
  - Burg PSD versus max power/mean power.
  - Max power versus mean power.

2.3 Classification of MI-based EEG signals

The basic idea behind classification is to establish a relation between the input EEG data and the class, for example, imagination of hands or feet, to which it belongs. In this study, both linear and nonlinear classifiers such as SVM, PSO-based SVM, *k*-NN, and LDA based on the principle of supervised learning are used [31,32]. For analysis purpose, training feature vectors are reorganized according to Eq. (1.1).

2.3.1 Support vector machine

SVM classifier is a popularly used pattern recognition algorithm for classification of EEG signals [31]. Its parameters are further optimized using PSO technique in this work.

2.3.2 PSO-based SVM

Appropriate selection of kernel function is an important step to obtain a good classifier so that it can provide improved results. Therefore, to find an appropriate kernel function PSO technique has been applied to obtain the values for scaling function sigma and box constraint value for the soft margin [33]. Here, an RBF kernel has been used. A training sample is selected, and initialization of velocity and position of the particles is done with an appropriate selection of the values of C and regularization parameter. After training of the samples with these initialization values fitness function for each particle is calculated and the personal best and global best positions are adjusted according to particle fitness value. The position and velocity are updated iteratively. After the maximum iterations are completed the process is stopped and the optimal kernel parameters are used to retrain the SVM classifier with the training samples. This is how the SVM classifier for prediction is obtained. The flowchart of parameter optimization using PSO is given in Fig. 1.7.

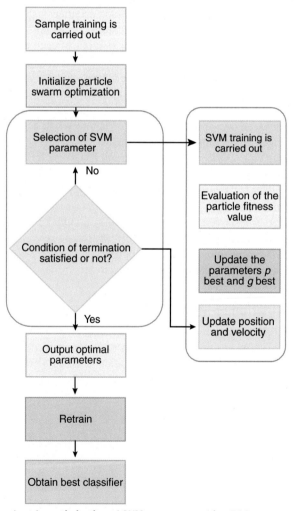

FIGURE 1.7 Flowchart for optimization of SVM parameters using PSO.

2.3.3   Linear discriminant analysis (LDA) and k-nearest neighbor (k-NN)

To further analyze the EEG signals classification has also been done using the most conventional classifiers, namely the LDA [10] and k-NN [19]. These algorithms have low computational complexity and are therefore feasible to be used for online BCI systems.

2.3.4   Extreme learning machine (ELM)

To test the accuracy of proposed method (PSO-SVM) comparison has been done to a popular classifying technique, that is, extreme learning machine (ELM). It simultaneously adjusts weights between input and hidden neurons and that between hidden neurons and output neurons with the help of a continuous probability density function (pdf) [21,34]. Its aim is to reduce the number of hidden neurons and get the minimum training error as given in Eq. (1.6).

$$min_{w,b,B} \left\| H\left(w_1, w_2 \ldots w_K, b_1, b_2, \ldots, b_K\right) B - T \right\|^2 \tag{1.6}$$

$$\text{where } T = \begin{bmatrix} t_1 \\ t_2 \\ . \\ . \\ t_K \end{bmatrix}$$

This algorithm fixes the input weights w_1 and bias b_1. The solution of the optimization problem 1 is stated as:

$$\tilde{B} = H'T \tag{1.7}$$

where H' is defined as the Moore–Penrose generalized inverse of the matrix of hidden layer given by H. Here H is defined as the hidden layer output matrix.

Sigmoidal activation function has been used in this work and probability density function is uniformly distributed from -1 to $+1$. The numbers of hidden neurons chosen are 4.

2.4   Analysis of different classifiers

For analyzing the performance of the classifier's classification accuracy is obtained which is the most important performance measure. To further validate the performance, other performances indices as given later are also computed.

2.4.1   Information transfer rate (ITR)

ITR is an important measure for evaluation of performance of the classifier measured in bits/trial. This is an important factor to be considered when the BCI systems are put to some real-time applications. It determines how fast is the transfer of signals to the device being controlled [35]. This transfer rate is highly dependent upon the number of imagery signals being generated and the type of classifier used. ITR denoted by B is given as:

$$B = \log_2\left(NP^P\right) + \left(1 - P\log_2\frac{1-P}{N-1}\right) \qquad (1.8)$$

Here, P is the classification accuracy of the classier and N is the number of MI tasks. ITR for all the classifiers is calculated for BP features.

2.4.2 Confusion matrix

For a more detailed evaluation of the classifier performance confusion matrix is obtained for each algorithm. It is a measure of the confusion of a classifier while predicting the class labels [36]. Parameters such as accuracy, precision, recall, specificity, sensitivity, F-score, geometric mean (g-mean), and Cohen's kappa are evaluated using it. These parameters are briefly explained in Table 1.1.

Table 1.1   Evaluation metrics for classifier performance.

Evaluation parameter	Formula
Accuracy	$\dfrac{Predicted\ classlabels - actual\ classlabels}{Total\ observations} \times 100$
Recall	$\dfrac{TP}{TP + TN}$
Precision	$\dfrac{TP}{TP + FP}$
Specificity	$\dfrac{TN}{FP + TN}$
Sensitivity	$\dfrac{TP}{TP + FN}$
F-score	$\dfrac{2 \times recall \times precision}{Recall + precision}$
g-mean	$\sqrt{Precision \times recall}$
Kappa	$\dfrac{P_o - P_e}{1 - P_e}$ P_o is the probability of correct classification and P_e is the hypothetical probability of chance agreement during classification

3   Results from case studies

All the results are evaluated for the validation of impending proposed classifier on four different benchmarks, since the experiments are conducted in three different steps which include preprocessing, feature extraction, and classification. So, the performance of each step of experiment is endorsed through various performance measures. In preprocessing, SNR is calculated before and after filtration using the spatial filter. In feature extraction evaluation of best nonlinear features using Friedman test, *t*-test is also used for ERD/ERS. In classification, classifier accuracy and different performance measures like kappa coefficients, sensitivity, and specificity are also evaluated separately for validating the proposed classifier; ITR is also calculated.

3.1   Preprocessing

A sixth-order IIR Butterworth filter is implemented for BP filtration to obtain the desired frequency range, that is, alpha (8–13 Hz) band and beta (14–30 Hz) band. For removal of unwanted noise and signals an SSL filter has been used. PSD, signal to noise ratio (SNR), and variance for all the subjects and classes were calculated before and after the filtration using SSL filter. Results for Class 1 (LH) observations for all the subjects are demonstrated in Table 1.2.

3.2   Feature extraction for automated classification

Features of different classes are obtained using seven different feature extraction methods. One feature per trial is calculated for all the subjects. Further, to test the efficacy of all the feature extraction methods two statistical tests, namely *t*-test and Friedman test, were conducted. *t*-test is applied on observations of each class to compare the data from two electrodes with the help of which quantification of ERD/ERS phenomenon is validated. On testing it is observed that the null hypothesis for each class is rejected. The results are demonstrated in Table 1.3. Similarly, Friedman test is conducted upon subject K3b and all classes taking into consideration combination of two different features at a time. *p*-value and chi-square value are evaluated for each feature combination. In Class 1 (LH) mean power features when com-

Table 1.2   PSD, SNR, and variance for subjects K3b, K6b, and L1b.

Subject	PSD before SSL filtering (dB)	PSD after SSL filtering (dB)	SNR before SSL filtering	SNR after SSL filtering	Variance before SSL filtering	Variance after SSL filtering
K3b	−22.4092	−51.8508	9.6934	0.0332	60.3564	21.536
K6b	−20.4585	−50.5021	7.6179	0.0082	54.986	15.987
L1b	−23.9875	−56.8734	5.9699	0.0017	51.982	24.274

Table 1.3 *p*-values obtained for different electrode pair combination for K3B, K6b, and L1b subjects.

Subject	Electrode pair (*p*-value)		
	C3-Cz	C3-C4	C4-Cz
K3b	0.0986	0.0831	0.0642
K6b	0.0865	0.0653	0.0742
L1b	0.0653	0.0732	0.0521

pared with Burg PSD and Welch PSD features reported a value of $\chi^2 = 157.1631$ and $p = 4.72e^{-36}$. Yule–Walker PSD and periodogram PSD reported a value $\chi^2 = 242.517$ and $p = 1.11e^{-54}$ with respect to mean power. Mean power, periodogram, and BP features proved to be the best choice of features and hence only these features are considered for further classification. The results are demonstrated in Table 1.4. RH and foot data results were also in coherence with the aforementioned results and hence these three features were further processed along with the nonlinear features.

3.3 Classifiers

Binary-class classification is performed using all the features. The feature vectors corresponding to two different classes are taken into account simultaneously. These are given as an input to the classifier. After the classifier is trained, the effectiveness of training of the classifier is tested with the help of test dataset. Different training techniques are applied for all classifiers. In *k*-NN classifier value of *k* is set to 5 and Euclidean distance calculated between training and testing data points and the nearest *k* neighbors. Voting technique is used to predict the class labels. For LDA the discriminant hyperplane parameters are computed for all the combinations considered for classification. In SVM, training was done using RBF kernel and the scaling factor sigma is kept 1. The training set was trained using a tenfold cross-validation scheme and the *C* and regularization were evaluated. These parameters were optimized with the help of PSO. The mentioned parameters were randomly generated as particles. The velocity and position of the particle are updated accordingly to maximize the cost function which in this case is classification accuracy. The optimization using PSO was conducted for all the three subjects and we obtained different values for *C* and regularization. The values of *C* and regularization parameter for subject K3b are 10.234 and $3.1674e^{-5}$, for K6b are 11.543 and 2.2684_e^{-4}, and for L1b are 7.765 and $4.1824e^{-4}$. The continuous iteration process gradually tends to maximal fitness to provide the optimum parameters.

The classification results obtained for the three linear features and all the subjects are tabulated in Tables 1.5, 1.6. Table 1.7 demonstrates the accuracy obtained using nonlinear features. Total accuracy for a binary combination and individual classification accuracy of the MI signals are depicted. It is observed that best classification accuracy for BP features is obtained in case of linear features. Accuracy for K3b

Table 1.4 Friedman's test result for evaluation of best features tested at a significance level of $p = 0.05$.

MI	Feature extraction method	Max power (p, χ^2)	Yule–Walker PSD (p, χ^2)	Burg PSD (p, χ^2)	Welch PSD (p, χ^2)	Periodogram PSD (p, χ^2)	Band power (p, χ^2)
LH	Mean power	$2.3e^{-46}$, 204.3	$1.11e^{-54}$, 242.5	$4.72e^{-36}$, 157.1	$4.72e^{-36}$, 157.16	$4.57e^{-55}$, 244.2	$3.87e^{-395}$, 106.2
	Max power	—	$4.49e^{-11}$, 43.3	0.01313, 2.2	$2.17e^{-11}$, 44.8	$8.61e^{-17}$, 69.26	$2.57e^{-5}$, 20.7
	Yule–Walker PSD	—	—	$3.07e^{-4}$, 13.02	0.9952, $3.68e^{-5}$	$7.56e^{-4}$, 11.34	0.0014, 10.5
	Burg PSD	—	—	—	0.0002, 13.6	$3.76e^{-11}$, 43.7	0.000065, 1.8
RH	Mean power	$2.17e^{-46}$, 204.5	$1.11e^{-53}$, 237.94	$5.49e^{-33}$, 143.14	$2.66e^{-53}$, 236.2	$4.31e^{-54}$, 239.82	$2.59e^{-21}$, 21.98
	Max power	—	$9.27e^{-10}$, 37.47	0.1947, 1.68	$3.92e^{-11}$, 43.65	$4.73e^{-16}$, 65.91	$1.17e^{-02}$, 1.64
	Yule–Walker PSD	—	—	0.0072, 7.22	0.4057, 0.69	0.0002, 14.04	$8.67e^{-5}$, 19.82
	Burg PSD	—	—	—	0.0001, 10.81	$2.36e^{-10}$, 40.14	0.00072, 8
Foot	Mean power	$2.03e^{-36}$, 158.8	$1.22e^{-44}$, 196.4	$2.53e^{-34}$, 149.25	$3.46e^{-46}$, 203.58	$7.34e^{-46}$, 202.08	$1.57e^{-20}$, 100.21
	Max power	—	$5.93e^{-10}$, 38.3	0.0107, 6.5	$5.49e^{-10}$, 38.5	$5.89e^{-12}$, 47.36	0.00216, 2.25
	Yule–Walker PSD	—	—	0.0006, 11.8	0.7447, 0.11	0.0028, 8.9	0.0031, 6.77
	Burg PSD	—	—	—	0.0005, 12.2	$9.68e^{-8}$, 28.4	0.0654, 10.56

—, no comparison between similar features.

subject using LDA classifier on the three combinations comes out to be 88%, 87%, and 78%. The nonlinear features clearly show a better classification accuracy which comes out to be 89.01%, 94.08%, 91.66% for subject K3b using LDA classifier. Therefore, to make linear features competent with the nonlinear ones the linear features having the best classification accuracy are selected to be tested with PSO-SVM.

Table 1.8 shows the classification results obtained using SVM-PSO and ELM classifiers along with their computational time both without normalizing BP features and with normalizing BP features. To further increase the chances of their enhancement of accuracy BP features have been normalized by obtaining the z-score of the data having 0 mean and standard deviation 1. The proposed SVM-PSO classifier has been compared with ELM so as to validate the results in accordance with the latest classifier that are being used nowadays. It can be observed that for subject K3b the accuracy for the three combinations has shown a considerable increase. Also it can

Table 1.5 Classification accuracy (%) for linear band power features for three subjects and various classifiers.

Subject	Classifier	LH versus RH			RH versus foot			LH versus foot		
		LH	RH	Average	RH	Foot	Average	LH	Foot	Average
K3b	k-NN	80	82	81.11	100	86.67	93.33	100	71.11	85.56
	SVM	97.78	78	87.78	100	77.78	88.88	100	71.11	85.86
	LDA	88.89	88.89	88.88	100	73	87	100	57.78	78.88
K6b	k-NN	43.33	50	46.67	70	50	60	63.33	50	53.33
	SVM	66.67	50	58.33	87	36.67	61.67	83.33	37	60
	LDA	63.33	43.33	53.33	57	43.33	50	60	43.77	51.67
L1b	k-NN	46.67	66.67	56.67	70	50	60	56.67	26.67	53.33
	SVM	50	53	51.67	53.33	63	58	57	50	53.33
	LDA	46.67	73	61	53.33	60	56.67	73	36.67	55

Table 1.6 Classification accuracy (%) for linear mean power and periodogram PSD features for three subjects and various classifiers.

Linear feature	Subject	Classifier	LH versus RH			RH versus foot			LH versus foot		
			LH	RH	Average	RH	Foot	Average	LH	Foot	Average
Mean power	K3b	k-NN	55.56	60	57.78	60	56.67	58.33	73.33	46.67	59.44
		SVM	24.44	91.11	57.78	93.33	37.78	65.56	93.33	25.56	59.44
		LDA	85.55	40	62.778	83.33	25.56	63.88	91.11	32.22	61.66
	K6b	k-NN	53.33	38.33	46.67	46.67	30	38.33	55	55	55
		SVM	58.33	33.33	45.83	56.67	38.33	47.5	40	68.33	54.17
		LDA	63.33	30	46.66	46.67	40	43.33	61.67	40	50.83
	L1b	k-NN	41.67	45	44.17	53.33	58.33	55.83	50	55	52.5
		SVM	78.33	10	44.17	16.67	81.67	49.17	46.67	56.67	51.67
		LDA	96.67	3	50	45	48.33	46.67	68.33	28.33	47.5
Periodo-gram PSD	K3b	k-NN	50	48.89	49.44	63.33	65.56	64.44	66.67	50	58.33
		SVM	55.56	14.44	96.67	85.56	32.22	58.89	95.56	20	57.78
		LDA	40	90	65	85.56	40	62.77	94.44	30	62.22
	K6b	k-NN	48.33	40	44.17	51.67	58.33	55	41.67	63.33	52.5
		SVM	78.33	16.67	47.5	3	93.33	48.33	65	33.33	49.17
		LDA	71.67	20	45.83	93.33	0	46.66	23.33	58.33	40.83
	L1b	k-NN	50	50	50	58.3	46.67	52.55	48.33	41.67	45
		SVM	25	75	50	50	48.33	49.17	90	15	52.5
		LDA	38.33	58.33	48.33	56.67	48.33	52.55	73.33	43.33	58.33

Table 1.7 Classification accuracy (%) for correlation coefficient and approximate entropy nonlinear features for three subjects and various classifiers.

Nonlinear feature	Subject	Classifier	LH versus RH			RH versus foot			LH versus foot		
			LH	RH	Average	RH	Foot	Average	LH	Foot	Average
Correlation coefficient	K3b	k-NN	59.56	65.21	67.78	69.23	76.88	68.93	85.43	76.69	82.43
		SVM	54.24	93.89	77.88	83.53	77.79	75.53	83.43	85.06	81.94
		LDA	88.89	89.23	89.01	93.33	95.36	94.08	91.31	88.22	91.66
	K6b	k-NN	73.97	49.83	66.87	54.97	60.32	58.30	56.76	72.98	65.65
		SVM	68.93	65.33	66.93	76.67	83.13	79.51	49.78	58.32	54.87
		LDA	73.93	80.34	76.69	61.35	54.23	57.93	71.67	69.56	70.83
	L1b	k-NN	51.66	78.34	64.77	63.92	68.90	65.93	87.65	79.76	82.59
		SVM	74.65	75.66	74.27	76.89	81.29	80.87	46.87	59.97	55.67
		LDA	86.97	65.39	75.32	55	58.33	56.69	87.23	78.93	85.35
Approximate entropy	K3b	k-NN	58.34	49.59	55.54	69.63	75.56	73.54	96.97	30.21	68.23
		SVM	67.96	64.39	66.17	95.46	89.23	92.74	59.86	50	56.79
		LDA	47.23	92.89	70.34	75.26	70.45	72.97	74.84	50.98	63.21
	K6b	k-NN	40.43	58.32	50.17	61.69	68.93	65.13	59.57	63.33	62.5
		SVM	82.93	88.98	86.51	63	83.53	78.83	65	63.33	62.17
		LDA	69.79	65	67.83	73.63	84.34	80.56	43.83	45.03	42.83
	L1b	k-NN	61.23	64.29	63.21	68.43	66.69	52.55	48.33	51.67	49.56
		SVM	61.98	64.38	62.98	59.43	78.23	49.17	90.87	45	72.5
		LDA	77.43	82.33	80.93	86.67	88.43	87.55	73.33	78.43	75.33

Table 1.8 Classification accuracy (%) of linear band power features without normalization and with normalization using SVM-PSO and ELM classifier.

Feature type	Subject	Classifier	LH versus RH		RH versus foot		LH versus foot	
			Accuracy (%)	Training time (s)	Accuracy (%)	Training time (s)	Accuracy (%)	Training time (s)
Without normalization	K3b	ELM	90	0.2496	93.33	0.156	87.81	0.624
		SVM-PSO	94.44	0.563	91.11	0.457	90	0.768
	K6b	ELM	55	0.312	48.33	0.312	46.67	0.2028
		SVM-PSO	60.7	0.452	62.67	0.654	65.55	0.765
	L1b	ELM	51.67	0.936	48.33	0.624	55	0.312
		SVM-PSO	64.4	0.1256	64.67	0.675	58.33	0.764
With normalization	K3b	ELM	91	0.158	91.36	0.106	88.81	0.514
		SVM-PSO	94.55	0.463	90.11	0.367	90.98	0.838
	K6b	ELM	56.89	0.234	50.33	0.3452	47.77	0.108
		SVM-PSO	61.78	0.562	64.67	0.546	69.55	0.663
	L1b	ELM	52.27	0.736	50.33	0.528	56.89	0.248
		SVM-PSO	66.43	0.3436	67.67	0.595	60.87	0.587

be seen that accuracy obtained using both the classifiers is nearly the same but the computation time varies significantly from each other.

In order to quantify the discriminative power of SVM-PSO and ELM classifiers, significant difference has been tested between the classification's accuracies obtained using these two classifiers with the help of t-test that can be termed as ranking of classifiers. The null hypothesis ($h = 1$ and $p = 0.069$) was rejected suggesting that there is a significant difference between the two classifiers. The probability of obtaining classification accuracy above 90% for subject K3b is 0.342 and 0.254 using SVM-PSO and ELM, respectively. Similarly, for subjects K6b and L1b the probability to get accuracy above 60% is 0.453 and 0.167 for SVM-PSO and ELM, respectively.

To check the robustness of classifiers intersubject classification is performed. The training of classifier is done using the features of all the three subjects and then the classifier is tested with the test data of any one of the subjects. The results are as depicted in Table 1.9.

Performance measures calculated with the help of the confusion matrix are depicted in Table 1.10. Average values for Cohen kappa, sensitivity, specificity, precision, recall, g-mean, and F-score are obtained for all the subjects and all the classifiers for all feature extraction techniques. But only the result for BP feature extraction method is presented in this chapter. The values of kappa, sensitivity, specificity, precision, recall, F-score, and g-mean are 0.762, 0.941, 0.899, 0.912, 0.898, 0.920, and 0.97, respectively, for subject K3b is obtained for PSO-SVM classifier.

The results in Table 1.11 depict the ITR calculated for all the classifiers and subjects corresponding to all MI signal combination.

4 Discussion

The detection of best MI EEG features requires a lot of research and inspection. Selection of best features from a pool of time–frequency-based features requires rigorous computations in different aspects such as the computation cost and difference in properties of the features. Similarly, for choosing the best classifier a lot of thought goes behind in improving the existing techniques and making it more efficient for the

Table 1.9 Intersubject classification accuracy (%) while taking training data of all the subjects and test data of any one subject at a time using SVM-PSO classifier.

Test data/ Accuracy	LH versus RH			RH versus foot			LH versus foot		
Subject	**LH**	**RH**	**Total**	**RH**	**Foot**	**Total**	**LH**	**Foot**	**Total**
K3b	93.33	75.56	84.44	100	87	93	100	71.11	85.56
K6b	76.67	20	48.33	80	43	62	83	20	51.67
L1b	23.33	83.33	53.33	56	93	56	47	70	58.33

Table 1.10 Statistical average performance measures for different classifiers.

Subject	Classifier	Kappa	Sensitivity	Specificity	Precision	Recall	F-score	g-mean
K3b	k-NN	0.654	0.8	0.822	0.818	0.8	0.808	0.811
	LDA	0.458	0.772	0.888	0.888	0.888	0.888	0.888
	SVM	0.432	0.678	0.456	0.567	0.124	0.668	0.567
	PSO-SVM	0.762	0.941	0.899	0.912	0.898	0.920	0.977
K6b	k-NN	0.653	0.736	0.564	0.234	0.735	0.543	0.938
	LDA	0.569	0.678	0.764	0.456	0.761	0.875	0.871
	SVM	0.349	0.189	0.764	0.761	0.901	0.456	0.236
	PSO-SVM	0.671	0.750	0.823	0.793	0.951	0.894	0.946
L1b	k-NN	0.418	0.678	0.456	0.224	0.446	0.2123	0.876
	LDA	0.738	0.345	0.789	0.872	0.889	0.2319	0.563
	SVM	0.3452	0.678	0.671	0.563	0.098	0.456	0.89
	PSO-SVM	0.782	0.689	0.810	0.890	0.9	0.531	0.865

Table 1.11 Information transfer rate of all the classifiers and three subjects.

Subject	LH versus RH				RH versus foot				LH versus foot			
	k-NN	SVM	LDA	SVM-PSO	k-NN	SVM	LDA	SVM-PSO	k-NN	SVM	LDA	SVM-PSO
K3b	0.464	0.496	0.03	0.470	0.470	0.53	0.645	0.442	0.412	0.255	0.404	0.404
K6b	0.008	0.039	0.032	0.012	0.039	0.01	0.029	0.039	0.029	0.008	0.003	0.007
L1b	0.464	0.496	0.012	0.008	0.018	0.01	0.029	0.001	0.003	0.007	0.003	0.020

purpose of classification. Taking care of the features extracted and classifier used is a crucial step because translation of the EEG signals from offline to online application depends very much on these two factors. Therefore, a special emphasis has been given in this chapter for the same.

Various feature extraction techniques are used and their effectiveness for successful discrimination of MI is tested. The features used are mostly time–frequency-based because they are considered to be most efficient in finding MI EEG signals. Also, extraction of the signal power-based feature proves to be appropriate in this scenario because ERD/ERS phenomenon shows changes in power level of the signal in different parts of the brain. This phenomenon has tried to be proven with the help of statistical analysis, which has not been done previously by researchers [37]. In lieu of this, a t-test has been conducted on both the electrodes of a class to validate this phenomenon. Hypothesis testing ensured the difference or similarity in the signals obtained from any of the electrodes simultaneously. In case of the features used Friedman test has been used for finding the similarity or correlation between any two features. It was observed that there is an interrelation between mean power with all other PSD features and max power with all other PSD features for all the classes. As analysis has also been done on nonlinear features, it is observed that they significantly show an improvement in classification accuracy over the regular linear features. Nonlinear dynamics of different biological signals have been documented to be functionally valid. A comparison has been made as to how the linearity and nonlinearity play a role in achieving the desired classification accuracy. After successful feature evaluation, they were fed to the classifier.

The best classifier for a particular subject cannot be determined. It varies from subject to subject and feature to feature. But in general, it is observed that LDA classifier outperformed all other classifiers in most of the cases. However, the parameters of a classifier can be played with to get a better output. Further, as BP features were categorized as the best features, PSO is used to enhance the performance of the SVM. Finding of appropriate kernel and box constraint value to be used with SVM classifier proved to be a rigorous task and requires thorough analysis and several rounds of classification to find the best parameter values. The main motivation behind using SVM is that it converts the vector into a higher dimensional space and constructs a hyperplane that separates the datas. ELM classifier also required proper selection of activation function and its threshold value. The PSO-SVM classifier performs better than ELM classifier in terms of accuracy (94%) although the computation time of it is more. The conduction of t-test on the results signifies that SVM-PSO mostly yields a higher accuracy. The probability of obtaining accuracy above 90% by SVM-PSO for K3b subject is 0.342, whereas for ELM it is 0.254. On further observation it can be seen that when MI signals for LH and RH are used, they mostly yield the highest classification accuracy. It can also be seen that normalization of features vectors can lead to a slight increase in the classification accuracy in some of the instances. The performance of the proposed classification technique with other international submissions [38,39] is comparable (92% for K3b) and in some cases better (94% for K3b) using linear feature extraction methods. The results show a drastic improvement in accuracy of classification after normalization and using PSO-SVM and ELM classifiers which is far more than that obtained by nonlinear feature extraction techniques proving that linear methods are also competent enough to be used after some processing.

The dataset used is based for offline analysis only. Real-time data need to be recorded and worked upon immediately while incorporating the aforementioned processes on an instantaneous basis to check the effectiveness. But this work is beyond the scope of the analysis presented in this chapter. Therefore, the proposed algorithms are not yet utilized for clinical applications. The results are only proposed for real-time applications. In future these algorithms can be tested for actual implementation under clinical supervision, for instance, for moving a wheelchair, movement of prosthetic arms, or using thoughts to move a cursor on a computer screen along with other neuro-aid applications involving some use of sensors [40,41].

5   Conclusion

The main objective of the thesis was to analyze the existing used for detection of MI signals and propose a better methodology to improve the earlier techniques. Here, an optimized version of the SVM classifier is suggested which in conjugation with BP features proves to be the best for imagery signal classification. Design of SVM classifier is simple but the main challenge is to select the best parameters for its design. With the help of PSO, classifier accuracy showed a tremendous improvement. Using SVM and BP as feature extraction technique accuracy for LH versus RH classification for K3b subject is reported to be 87.78%, whereas by using PSO-SVM for classification the accuracy increased to 94.44%. ELM reported an accuracy of 905 for the same. Similarly, for foot classification, there was an increase of accuracy which was as high as 5%. Hence, introduction of PSO proved to be of great advantage. There is still a scope of improvement in the present work on a wide front. A more efficient preprocessing filter such as CSP filter can be used for data averaging and artifact removal. As EEG signals are inherently nonlinear in nature, various new and unexplored feature extraction techniques such as fractal analysis can be done. The enhancement of classification accuracy using a multi-class framework for MI signals using suitable features in combination with effective classification algorithm seems to be promising for neuro-aid applications which may include moving a wheelchair or a robotic arm. Optimization in parameters of all the classifiers can be done using different optimization techniques. As there is a boom of deep learning classification algorithms these days, they can be put to use instead of the conventional algorithms to get better classification results that can be used for real-time applications such as device control. Further, these signals in conjugation with some hardware systems such as motors can be used for various neuro-aid applications like wheelchair control or prosthetic arm movement control.

Instead of using the EEG signals corresponding to MI, other signals such as those related to emotions can be extracted and processed using the same methodology.

Through this work, we have tried to contribute a little toward the BCI systems improvement.

To sum up, major conclusions drawn from the presented work are as follows:

- From the detailed analysis of various preprocessing techniques, it is concluded that Butterworth filter as a BP filter and SSL filter as a spatial filter are the best for alpha and beta wave extraction and signal averaging and cleaning, respectively.
- BP features and PSO-based SVM are the best pair MI signal classification and processing.
- ELM classifier has a potential to be used as a classification technique for BCI applications.

Acknowledgment

The authors would like to acknowledge with much obligation the crucial role of Graz University and the team who involves in EEG signal recordings and makes it available online for new findings and analysis.

6   Compliance with ethical standards

The analysis has been done on the downloaded BCI competition data. No human or animal subjects were involved in this study.

7   Conflict of interest

The authors declare there is no conflict of interest with regards to this submission.

References

[1] B. Rebsamen, C. Guan, H. Zhang, C. Wang, C. Teo, M.H. Ang, E. Burdet, A brain controlled wheelchair to navigate in familiar environments, IEEE Trans. Neural Syst. Rehabil. Eng. 18 (2010) 590–598.
[2] M.R. Williams, R.F. Kirsch, Evaluation of head orientation and neck muscle EMG signals as command inputs to a human–computer interface for individuals with high tetraplegia, IEEE Trans. Neural Syst. Rehabil. Eng. 16 (2008) 485–496.
[3] D. Coyle, J. Garcia, A.R. Satti, T.M. McGinnity, EEG-based continuous control of a game using a 3 channel motor imagery BCI: BCI game, Computational Intelligence, Cognitive Algorithms, Mind, and Brain (CCMB), IEEE, Paris, France, 2011, pp. 1–7.
[4] A. Schlögl, F. Lee, H. Bischof, G. Pfurtscheller, Characterization of four-class motor imagery EEG data for the BCI-competition 2005, J. Neural Eng. 2 (2005) L14.

[5] T. Wang, J. Deng, B. He, Classifying EEG-based motor imagery tasks by means of time–frequency synthesized spatial patterns, Clin. Neurophysiol. 115 (2004) 2744–2753.

[6] G. Pfurtscheller, C. Brunner, A. Schlögl, F.L. Da Silva, Mu rhythm (de) synchronization and EEG single-trial classification of different motor imagery tasks, NeuroImage 31 (2006) 153–159.

[7] X. Yong, C. Menon, EEG classification of different imaginary movements within the same limb, PloS One 10 (2015) 0121896.

[8] J. Atkinson, D. Campos, Improving BCI-based emotion recognition by combining EEG feature selection and kernel classifiers, Expert Syst. Appl. 47 (2016) 35–41.

[9] M. Murugappan, N. Ramachandran, Y. Sazali, Classification of human emotion from EEG using discrete wavelet transform, J. Biomed. Sci. Eng. 3 (2010) 390.

[10] N. Yalcin, G. Tezel, C. Karakuzu, Epilepsy diagnosis using artificial neural network learned by PSO, Turk. J. Electr. Eng. Comp. Sci. 23 (2015) 421–432.

[11] T. Zhang, W. Chen, LMD based features for the automatic seizure detection of EEG signals using SVM, IEEE Trans. Neural Syst. Rehabil. Eng. 25 (2017) 1100–1108.

[12] H. Yuan, A.J. Doud, A. Gururajan, B. He, Cortical imaging of event-related (de) synchronization during online control of brain-computer interface using minimum-norm estimates in frequency domain, IEEE Trans. Neural Syst. Rehabil. Eng. 16 (2008) 425.

[13] D. Gutierrez, Using single/multi-channel energy transform as preprocessing tool for magnetoencephalographic data-based applications, 2010 20th International Conference on Electronics, Communications and Computer (CONIELECOMP), IEEE, Cholula, Mexico, 2010, pp. 114–118.

[14] D.J. McFarland, L.M. McCane, S.V. David, J.R. Wolpaw, Spatial filter selection for EEG-based communication, Electroencephalogr. Clin. Neurophysiol. 103 (1997) 386–394.

[15] Z.J. Koles, M.S. Lazar, S.Z. Zhou, Spatial patterns underlying population differences in the background EEG, Brain Topogr. 2 (1990) 275–284.

[16] G. Townsend, B. Graimann, G. Pfurtscheller, A comparison of common spatial patterns with complex band power features in a four-class BCI experiment, IEEE Trans. Biomed. Eng. 53 (2006) 642–651.

[17] P. Herman, G. Prasad, T.M. McGinnity, D. Coyle, Comparative analysis of spectral approaches to feature extraction for EEG-based motor imagery classification, IEEE Trans. Neural Syst. Rehabil. Eng. 16 (2008) 317–326.

[18] M.C. Corsi, M. Chavez, D. Schwartz, L. Hugueville, A.N. Khambhati, D.S. Bassett, F. De Vico Fallani, Integrating EEG and MEG signals to improve motor imagery classification in brain–computer interface, Int. J. Neural Syst. 29 (1) (2019) 1850014.

[19] J. Kevric, A. Subasi, Comparison of signal decomposition methods in classification of EEG signals for motor-imagery BCI system, Biomed. Signal Process. Control 31 (2017) 398–406.

[20] F. Lotte, M. Congedo, A. Lécuyer, F. Lamarche, B. Arnaldi, A review of classification algorithms for EEG-based brain–computer interfaces, J. Neural Eng. 4 (2007) R1.

[21] Y. Ma, X. Ding, Q. She, Z. Luo, T. Potter, Y. Zhang, Classification of motor imagery EEG signals with support vector machines and particle swarm optimization, Comput. Math. Methods Med. 2016 (2016) 1–8.

[22] G.B. Huang, Q.Y. Zhu, C.K. Siew, Extreme learning machine: theory and applications, Neurocomputing 70 (2006) 489–501.

[23] B. Blankertz, K.R. Muller, D.J. Krusienski, G. Schalk, J.R. Wolpaw, A. Schlogl, G. Pfurtscheller, J.R. Millan, M. Schroder, N. Birbaumer, The BCI competition III: validating alternative approaches to actual BCI problems, IEEE Trans. Neural Syst. Rehabil. Eng. 14 (2006) 153–159.

[24] F. Lee, R. Scherer, R. Leeb, C. Neuper, H. Bischof, G. Pfurtscheller, A comparative analysis of multi-class EEG classification for brain computer interface, Proceedings of the 10th Computer Vision Winter Workshop, Technical University of Graz, Austria, 2005, pp. 195–204.

[25] C. Neuper, R. Scherer, S. Wriessnegger, G. Pfurtscheller, Motor imagery and action observation: modulation of sensorimotor brain rhythms during mental control of a brain–computer interface, Clin. Neurophysiol. 120 (2009) 239–247.

[26] A. Widmann, E. Schröger, Filter effects and filter artifacts in the analysis of electrophysiological data, Front. Psychol. 3 (2012) 233.

[27] T.C. Ferree, Spherical splines and average referencing in scalp electroencephalography, Brain Topogr. 19 (2006) 43–52.

[28] B. He, J. Lian, G. Li, High-resolution EEG: a new realistic geometry spline Laplacian estimation technique, Clin. Neurophysiol. 112 (2001) 845–852.

[29] A. Widmann, E. Schröger, B. Maess, Digital filter design for electrophysiological data—a practical approach, J. Neurosci. Methods 250 (2015) 34–46.

[30] C. Kim, J. Sun, D. Liu, Q. Wang, S. Paek, An effective feature extraction method by power spectral density of EEG signal for 2-class motor imagery-based BCI, Med. Biol. Eng. Comput. 2 (2018) 1–14.

[31] N. Dey, A. Ashour (Eds.), Classification and Clustering in Biomedical Signal Processing, IGI Global, Hershey, PA, 2016.

[32] N. Dey, A.S. Ashour, S. Borra (Eds.), Classification in BioApps: Automation of Decision Making, vol. 26, Springer, Cham, 2017.

[33] S. Siuly, Y. Li, Improving the separability of motor imagery EEG signals using a cross correlation-based least square support vector machine for brain–computer interface, IEEE Trans. Neural Syst. Rehabil. Eng. 20 (2012) 526–538.

[34] T.N. Lal, M. Schroder, T. Hinterberger, J. Weston, M. Bogdan, N. Birbaumer, B. Scholkopf, Support vector channel selection in BCI, IEEE Trans. Biomed. Eng. 51 (2004) 1003–1010.

[35] H.I. Suk, S.W. Lee, A novel Bayesian framework for discriminative feature extraction in brain-computer interfaces, IEEE Trans. Pattern Anal. Mach. Intell. 35 (2013) 286–299.

[36] N.Y. Liang, P. Saratchandran, G.B. Huang, N. Sundararajan, Classification of mental tasks from EEG signals using extreme learning machine, Int. J. Neural Syst. 16 (2006) 29–38.

[37] S. Marsland "Machine learning: An algorithmic perspective" Boca Raton: Chapman & Hall/CRC 2009.

[38] B.J. Edelman, B. Baxter, B. He, EEG source imaging enhances the decoding of complex right-hand motor imagery tasks, IEEE Trans. Biomed. Eng. 63 (2016) 4–14.

[39] B. Blankertz, Results of the BCI Competition III. Available from: http://www.bbci.de/competition/iii/results/bci_competition_iii_results_list.pdf, 2005.

[40] N. Dey, V. Bhateja, A.E. Hassanien, Medical Imaging in Clinical Applications, Springer International Publishing, Cham, 2016.

[41] C. Bhatt, N. Dey, A.S. Ashour (Eds.), Internet of Things and Big Data Technologies for Next Generation Healthcare, Springer, New York, 2017.

Biosensor and its implementation in diagnosis of infectious diseases

Susrita Sahoo[a], Satya Narayan Sahu[a], Subrat kumar Pattanayak[b],
Namrata Misra[a,c], Mrutyunjay Suar[a,c]

[a]School of Biotechnology, Kalinga Institute of Industrial Technology (KIIT), Deemed to be University, Bhubaneswar, India; [b]Department of Chemistry, National Institute of Technology, Raipur, India; [c]KIIT Technology Business Incubator, Kalinga Institute of Industrial Technology (KIIT), Deemed to be University, Bhubaneswar, India

1 Introduction

Over the past few years, biosensors have garnered widespread attention for being accurate, rapid, sensitive, and highly selective technology in the diagnosis of various infectious diseases at an affordable cost [1]. Following the invention of the first *true* oxygen electrode sensor by L.C. Clark for oxygen detection [2], there have been numerous advancements in sensitivity, selectivity, and multiplexing space of modern biosensors. In general, biosensor can be characterized as a compact analytical device including a biological or biologically acquired sensing element additionally combined inside or closely associated with a physicochemical transducer. The two basic operating principles of biosensor include *biological recognition* and *sensing* [3]. The major components of biosensors include a biological recognition system (bioreceptor), a transducer, and a signal processing system [4]. In spite of notable advancement in the field of health care: device and diagnostics over the past few years, detection of several infectious diseases is still a major challenge [6]. The widely used conventional laboratory-based diagnostics for the detection of various pathogens include nucleic-acid amplification, culture, immunoassays, and microscopy. However, majority of these experimental techniques have fundamental drawbacks. For instance, microscopy lacks sensitivity in many clinical series of development and culture is a time-consuming process. Similarly, immunoassays are highly sensitive, but labor intensive and challenging to perform multiplex detection. Likewise, nucleic-acid amplification tests (NAATs) such as polymerase chain reaction (PCR) offer molecular specificity but have complicated sample preparation methods and high chance of exhibiting false positives [5]. Vector-borne diseases are a crucial threat to human health globally, of which malaria is the most serious. Malaria is a fatal infectious disease transmitted to humans by mosquitoes and occurring mostly in the tropical countries [7]. According to *World Malaria Report*, released in November 2018, there were 219 million cases of malaria in 2017, up from 217 million cases in 2016. The estimated number of malaria deaths stood at 435,000 in 2017, which is quite alarming [8].

Computational studies employing various bioinformatics tools and methods help in understanding the underlying molecular details that would certainly aid in the development of biosensors and other biomedical diagnostics for detection of infectious disease. A comprehensive knowledge of binding mechanism at atomistic scale by molecular modeling and docking techniques is necessarily the preliminary work for biosensor designs. It has been reported that hemoglobin degradation/hemozoin formation are essential steps in the Plasmodium life cycle, and are promising targets for novel antimalarial discoveries. Herein, in silico approach employing homology modeling and docking method has been used to study the binding of heme with dihydrofolate reductase–thymidylate synthase (PfDHFR-TS) and heme detoxification protein (PfHDP), the two crucial validated targets for antifolate antimalarials. The resulting in-depth atomistic-level information of three-dimensional (3D) structure of PfDHFR-TS and PfHDP enzymes and the binding interactions with heme will provide functional insights for designing experimental framework to develop optimum biosensor models bearing the mutant enzymes to combat malaria.

1.1 Principle and components of a biosensor

The general aim of the design of biosensor is to enable rapid, convenient testing at the point of care where the sample was procured [72]. There are three so-called "generations" of biosensors as shown in Fig. 2.1. In the first-generation biosensors,

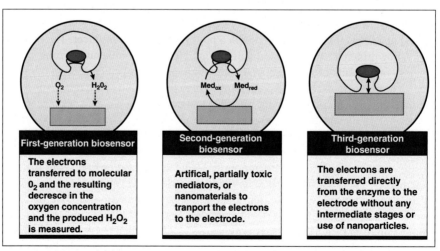

First-generation biosensor

The electrons transferred to molecular O_2 and the resulting decrese in the oxygen concentration and the produced H_2O_2 is measured.

Second-generation biosensor

Artifical, partially toxic mediators, or nanomaterials to tranport the electrons to the electrode.

Third-generation biosensor

The electrons are transferred directly from the enzyme to the electrode without any intermediate stages or use of nanoparticles.

FIGURE 2.1 Schematic representation of the working principle of the three major generations of biosensor.

the normal product of the reaction diffuses to the transducer and causes the electrical response, whereas the second-generation biosensors involve specific "mediators" between the reaction and the transducer in order to generate improved response, and in the third-generation biosensors, the reaction itself causes the response and no product or mediator diffusion is directly involved.

Biosensor can be defined as a scientific device comprising an immobilized biological material such as enzyme, antibody, nucleic acid, hormone, organelle, or whole cell, which can importantly interact with an analyte and produce physical, chemical, or electrical signals that can be measured. The operation of biosensor is generally based on the principle of signal transduction. The three major components widely used in biosensors are shown in Fig. 2.2.

The application of biosensor is utilized in various fields such as disease monitoring, drug discovery, and detection of pollutants, disease-causing microorganisms and markers that are indicators of a disease in bodily fluids (blood, urine, saliva, sweat) [8]. A typical biosensor consists of the following components (Fig. 2.3) [8].

1. *Analyte*: An analyte is a compound (e.g., glucose, urea, drug, pesticide) whose concentration has to be measured. Basically, biosensors include quantitative study of different substances by converting their biological actions into measurable signals. For instance, "histidine rich protein II" is an "analyte" in a biosensor designed to detect malaria [9].

2. *Bioreceptor*: It is a molecule that specifically recognizes the analyte. Enzymes, cells, aptamers, deoxyribonucleic acid (DNA), and antibodies act as bioreceptors in different biosensors. Biorecognition is process of generation of signal in the form of light, pH, mass change, or charge by the interaction of the bioreceptor with the analyte. There are multiple bioreceptors that are commonly used in the developed biosensors for malaria diagnosis, such as DNA aptamer, HRP-II specific antibodies, biotinylated DNA probe specific to Plasmodium [10,11,12].

3. *Transducer*: In a biosensor the function of transducer includes conversion of biorecognition event into a measurable signal. This method of energy conversion is known as signalization. Most transducers produce either optical or electrical signals that are usually proportional to the amount of analyte-bioreceptor interactions. Gold nanoparticle (GNPs) with poly-diallyl dimethyl ammonium chloride (PDDA) polymer and GNPs with poly(allylamine hydrochloride) (PAH) polymers, electrochemical impedance spectroscopy (EIS)

FIGURE 2.2 The major components of a functional biosensor.

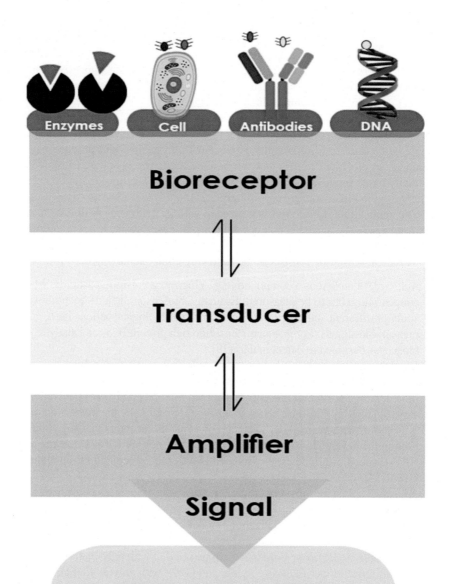

FIGURE 2.3 Schematic structure of a functional biosensor.

with electrode, and screen-printed gold electrodes functionalized with GNPs are some transducers used in aptamer-based biosensor, electrochemical-based biosensor, and immunosensors, respectively [10,13,14].

4. *Signal processing unit*: The signal processing unit includes the electronics of a biosensor that analyses the transduced signal and prepares it for display. It consists of complicated electronic circuit that performs signal conditioning such as amplification and conversion of signals from analog into the digital form. The processed signals are then quantified by the display unit of the biosensor. The display consists of a combination of hardware and software that produces output of the biosensor in a user-friendly manner [73]. The output signal can be numeric, graphic, tabular, or an image, depending on the requirements of the end user [74]. For instance, the output of biosensors for malaria diagnosis can be detected either by colorimetric assay, impedance response, optical measurement, electrochemical signal, or fluorescence [15,16,17,18].

1.2 Classification of biosensors

Biosensors can be classified according to their transducer type, the nature of their biological component, or bioreceptor [3,19] as shown in Fig. 2.4.

The bioreceptor constitutes the recognition system of a sensor toward the target analyte. It underpins molecular species that recognize analyte through biochemical mechanisms. They are involved in binding the analyte of interest to the sensor surface for the measurement. The receptor can be enzymes, antibody/antigen, nucleic acid/DNA, cellular structure, or biomimetic [3].

1. Enzymes are often used as bioreceptors because of their efficient binding capacity and catalytic activity. Updike and Hicks in 1967 reported the first enzyme-based

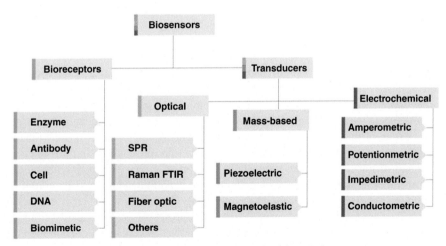

FIGURE 2.4 Schematic representation of general classification of biosensors.

biosensors [20]. Furthermore, enzymes being natural proteins are not consumed in the chemical reaction hence chosen as potential bioreceptor for sensor.

2. Immunosensors specialize antibody as bioreceptor to detect the particular antigen with high affinity [3].
3. The DNA sensors are generally designed based on the property that single-stranded nucleic acid binds to the complementary strands forming stable hydrogen bonds [21,22]. The first cell-based sensor was designed by Divies [23].

The transducer is a part of biosensors that converts biorecognition signal events into detectable signals; the later can be electrochemical (potentiometry, conductometry, impedimetry, amperometry, voltammetry), optical (colorimetric, fluorescence, luminescence, interferometry), calorimetric (thermistor), mass change (piezoelectric/acoustic wave), or magnetic in nature [3].

1. Electrochemical biosensors analyze the content of a biological sample by direct conversion of a biological event to an electronic signal. The first electrochemical impedance spectroscopy was reported in 1975 [24].
2. Optical biosensors are based on the optoelectronic techniques, and the transduction process gives rise to a change in the phase, amplitude, polarization, or frequency of the input light in response to the physical or chemical change produced by the biorecognition process [3].
3. On the contrary, magnetic biosensors detect magnetic micro and nanoparticles in microfluidic channels using the magneto-resistance effect and are observed to be very efficient in terms of sensitivity and size [25].

1.3 Comparison of current conventional lab techniques *vis-a-vis* biosensors for disease diagnosis

Well-timed precise diagnosis and inception of targeted antimicrobial treatment is important for an effective clinical management of infectious diseases [26,27,28,29,30,71]. Particularly, vector-borne diseases have become a humongous threat to tropical and developed countries because of continuous expose to deadly pathogens [31]. Furthermore, in many infectious diseases, the early phase symptoms are generally nonspecific, that is, clinically they cannot be differentiated from other infection, hence requiring definite laboratory information. Isolation of infections microorganisms is highly undesirable as a first-line approach in endemic situations, but despite of this it is the most direct and conclusive approach in medical diagnosis. There are many conventional techniques that are presently being practiced for medical diagnosis; however, each of these approaches has still major limitations. For instance, although serological techniques of isolation and direct detection of the pathogen are simpler, yet it might require many days until a sufficient number of antibodies are produced. Likewise, PCR-based processes are specifically complex, prone to contamination and require skilled manpower including bulky equipment for onsite applications; moreover, the samples need to be stored at very low temperatures, which is extremely challenging in rural settings [32]. Similarly, microarray technology is significantly costly to make it effective in a short time and valuable for point-of-care diagnosis. Against this backdrop, biosensors on contrary provide an easy-to-use, cost-effective, and sensitive platform that identifies pathogens rapidly and envisage effective treatment [26,27].

Additionally, unlike bioassays or conventional bioanalytical systems, biosensors do not require supplementary processing methods such as reagent addition, and are specified by permanent fixing of the assay design in the construction of the device [33]. The application of biosensors in detection of infectious disease is quite broad. It could be used for early detection of pathogens in blood [34] detection of virulence in vaccines, of dangerous pathogens present in environment, or in bandage like devices for treating septic wounds [35]. Conclusively, biosensors exhibit easier, faster, and cheaper results in comparison to traditional bioaffinity assays, also including high sensitivity and specificity of detection. The common conventional diagnostic methods for different infectious diseases are discussed in Table 2.1 [5].

1.4   Computational study of interaction of PfDHFR-TS and PfHDP with heme: promising biomarker targets for malaria

Among various vector-borne diseases, malaria is one of the prevalent virulent infectious diseases. It dominates over public health issue in more than hundreds of nations [50].

1.4.1   Current scenario of malaria diagnosis

Cost and benefit issues have impeded further worldwide availability of rapid diagnostic tests (RDTs) for malaria. Early stage diagnosis of malaria is essential to prevent the wide spread of malaria and will hence lead to reduced patient mortality through appropriate treatment.

1. Microscopic analysis
   Microscopic analysis is widely accepted as "gold standard" for diagnosis of malaria. Advantages of microscopic analysis include active turnaround time, an ability to determine species by thin smear, namely the prompt calculation of the percentage of infected red blood cells in the sample, economical, the preference of storing sample, and simple laboratory infrastructure. Various advanced modes of microscopy such as secondary speckle sensing microscopy and fluorescence in situ hybridization techniques have been lately reported as improved malaria diagnosis with a higher probability of detection [60,61].
2. Antibody-based RDTs
   RDTs are basically antibody-based immuno-chromatographic kits that detect either malaria species or malaria antigens. Advantages of RDT include analysis on the spot, capability to function without electricity, detection of multiple species, quick results, economical, the option of using unskilled or less-skilled technicians, and easy operation and interpretation.

Table 2.1 Laboratory diagnostic methods generally used for different infectious diseases.

Diseases	Site of infection	Sample	Diagnostic tests	Key results and challenges	References
Malaria	Liver	Blood, stool	Blood film microscopy, antibody-based rapid diagnostics, nucleic-acid amplification test	High diagnostic standard maintenance, lack of molecular targets for non-falciparum infections	[36,37,38]
AIDS	Immune systems	Blood, saliva, urine	CD4 T-cell counts, dipstick immunoassay, ELISA, Western bloat, viral load, nucleic-acid amplification test	Viral isolation and load determination are challenging; requires multiple biomarkers	[39,40,41]
Influenza virus infection	Respiratory tract	Nasal swab, sputum, blood	Hemagglutination-inhibition assay, ELISA, immunofluorescence assay, single radial hemolysis, culture, nucleic-acid amplification test	Strain-specific assays are essential due to the wide diversity of the virus	[42,43,44]
Dengue fever	Immune system	Blood	Culture, ELISA, nucleic-acid amplification test	Challenges in viral isolation, fluctuation in performance of diagnostic assay	[45,46]
Tuberculosis	Lung	Sputum, urine, blood	Sputum smear microscopy, IFN-release assay, tuberculin skin test, culture, nucleic-acid amplification test, ELISA	Lacks established biomarkers, challenging sample processing techniques	[47,48,49]

3. Nucleic-acid amplification tests

NAATs depend on the technique of detecting nucleic acid of the malaria parasite. NAATs are specifically recommended for epidemiological and research studies. Major advantages of NAATs involve efficient sensitivity, the capacity to detect one parasite per µL of blood sample, both qualitative and quantitative test, effective detection of multiple parasitic infections, the ability to process more than one sample at a time (high-throughput processing), and the ability to detect drug-resistant strains.

But all the diagnostic technique has following disadvantages:

1. Microscopic analysis: This off-site lab-based method requires personnel who are proficient in blood smear analysis and training of those personnel, and gives relatively slower results [62].
2. RDTs are not sensitive to low parasite density.
3. The disadvantages of NAATs are that they are time-consuming and expensive, have poor reproducibility at low concentrations, and require good laboratory facilities and skilled personnel [63].

Different species of *Plasmodium* like *Plasmodium vivax*, *P. falciparum*, *P. ovale*, and *P. malariae* primarily causes malaria in human. The most fatal species is *P. falciparum* which is transmitted through the bite of female *Anopheles* mosquito [51,52]. It causes organ failures and accumulates in the brain capillaries leading to coma in late stage. Out of the aforementioned species, *P. vivax* and *P. ovale* affect the liver after early stage of infection [51,52]. It is therefore important to choose drugs which are directly focused to the liver stage and ultimately cure the diseases [52]. When the liver cell erupts the merozoites attack the red blood cells. Within intraerythrocytic phase the merozoites enter through different forms. After then, daughter merozoites are formed. Those new species are responsible for attacking new red blood cells. Hence, the targeted drugs should minimize the infection of the disease [52]. In the wide range coverage of drug discovery field, both PfDHFR-TS and PfHDP are widely known targets for malaria [50,53,54]. *P. falciparum* proteins are showing amenable toward both liver and blood stage of human [50]. The identification of diseased markers and predicting their effects by computational approaches has potential to generate personalized tools for the diagnosis and treatment of malaria. Keeping this in view, herein, we are focusing on blood stage of the malaria parasite in order to provide computational insights that would be certainly useful in the development of heme sensing biomarkers of the *P. falciparum* (PfDHFR-TS and PfHDP) using homology modeling and docking approach. Homology modeling is the most commonly used computational approach for deciphering the 3D structure of a protein. The molecular docking is a computational method to calculate the binding affinity and interaction energy of receptor–ligand complex.

2   Materials and methods

2.1   Determination of three-dimensional structure of *PfHDP* protein

The tertiary structure of PfHDP protein was generated using Modeller *v* 9.21 program [58]. For template selection, BLASTp search [65] was performed against the

PDB database [66] in addition to I-TASSER tool [67] to find X-ray crystallographic structures with maximum identity and lower E-value. Finally, the crystal structure with PDB ID: 5NV6 was selected as template for modeling which exhibited a query coverage of 27%, and 77% identity. Out of 200 distinct models generated through Modeller program, the best model was selected based on the highest DOPE (Discrete Optimized Protein Energy) score which was further subjected to loop and side-chain refinement using GalaxySite [59] program.

2.2   Docking of PfDHFR-TS and heme

The crystal structure of PfDHFR-TS (PDB ID: 3UM6) and heme (Pub Chem Id: 26945) was docked using the AutoDock Vina software [56]. Grid box parameters were set in such a way so as to allow for a suitably sized cavity space large enough to accommodate each compound within the binding site of the proteins. The Lamarckian Genetic Algorithm was used to search for the best ligand conformer with other parameters set as default value. The final conformations were clustered and ranked according to the AutoDock scoring function. The docked conformations with the lowest binding affinity were selected for further analysis of the binding interface and interacting residues using PyMOL tool (http://www.pymol.org/) tool and Discovery studio visualizer 4.0 software [57].

2.3   Docking of PfHDP and heme

The modeled PfHDP-TS was docked with heme complex using the same protocol as followed for PfDHFR-TS–heme complex.

2.4   Structure validation of the modeled structures

The developed PfHDP model was validated stereochemically using various structure validation tools, namely PROCHECK [68], ProSA [55], VERIFY3D [69], and ERRAT [70] web servers. All these structure validation tools use disparate methods to calculate quantitative scores that can be used to assess model quality and guide further selection of the most accurate protein models. PROCHECK corroborates the stereochemical quality of protein structure by evaluating the accuracy of the dihedral angles (ϕ and ψ) in the Ramachandran plot. The ProSA calculates an overall Z-score that can be used to evaluate the structural quality of a computationally determined protein model with respect to crystal and NMR structures present in PDB. VERIFY3D is used to analyze the compatibility of the 3D protein model with its own amino-acid sequence. ERRAT outputs an overall model quality factor to examine nonbonded atomic interactions. A higher score from VERIFY3D and ERRAT indicates a stable reliable structure. Table 2.2 summarizes various computational tools and databases used in the present bioinformatics study.

Table 2.2   A brief overview of different computational tools and databases used in the present study.

Tools and databases	Description	References
BLAST	Finds regions of local similarity between sequences.	[65]
Protein Data Bank	Repository of 3D structures of large biological molecules including proteins and nucleic acids.	[66]
Modeller 9.19	Homology or comparative modeling of protein three-dimensional structures.	[58]
VERIFY3D	Determines the compatibility of an atomic model (3D) with its own amino-acid sequence (1D) by assigning a structural class based on its location and environment.	[69]
ERRAT	Analyzes the statistics of nonbonded interactions between different atom types and plots the value of the error function versus position of a 9-residue sliding window, calculated by a comparison with statistics from highly refined structures.	[70]
PROCHECK	Checks the stereochemical quality of a protein structure by analyzing residue-by-residue geometry and overall structure geometry.	[68]
ProSA	Calculates an overall quality score for a specific input structure. If this score is outside a range characteristic for native proteins the structure probably contains errors.	[55]
AutoDock Vina	Protein-ligand docking	[56]
PyMOL	Molecular visualization of biomolecule structures	[64]
Discovery Studio 4.0	Comprehensive software suite for analyzing and visualizing macromolecules.	[57]

3   Results and discussion

3.1   Binding of heme with PfHDP and PfDHFR-TS

The 3D structure of PfHDP was predicted using Modeller software. The modeled Pf-HDP protein (Fig. 2.5) exhibits six strands of antiparallel β-sheet and five α-helices.

The quality of the predicted 3D models was further assessed using various structure validation tools. The PROCHECK results for PfHDP (Fig. 2.6A) indicated the developed structure to possess reliable stereochemical quality having more than 73% of the residues in the allowed region of the Ramachandran plot. In order to detect potential errors in the structures, Z-score was calculated using ProSA-web

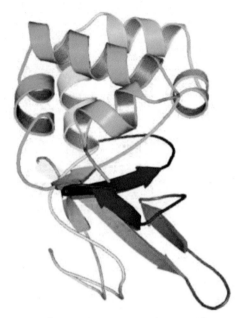

FIGURE 2.5 Ribbon representation of the overall molecular architecture of PfHDP.

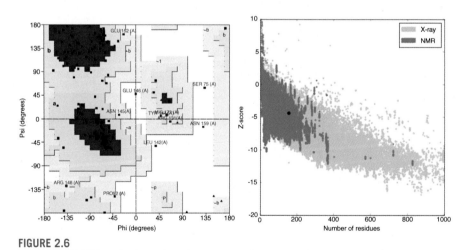

FIGURE 2.6

(A) Ramachandran plot for the PfHDP model generated using PROCHECK server. (B) ProSA-web Z-scores of all protein chains in PDB determined by X-ray crystallography (*light blue*) and NMR spectroscopy (*dark blue*) with respect to their length.

program. The score outside a range characteristic for native proteins indicates erroneous structures. In the present study, the Z-score of PfHDP falls within the range of experimentally determined proteins in PDB (Fig. 2.6B). Furthermore, the high-scoring values obtained using various other structure validation tools, namely VERI-

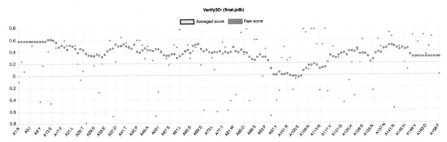

FIGURE 2.7 VERIFY3D results showing percentage of residues with average 3D-1D score >0.2.

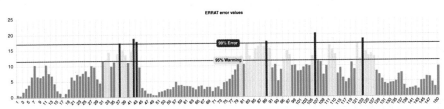

FIGURE 2.8 ERRAT score indicating percentage of the protein for which the calculated error value falls below the 95% rejection limit.

FY3D and ERRAT, demonstrated the overall accuracy of the modeled structures (Figs. 2.7 and 2.8).

Following structure determination, molecular docking of PfHDP and PfDHFR with heme was preformed to explore their potential binding interfaces. The binding interaction of PfHDP–heme and PfDHFR–heme docked complex is shown in Figs 2.9 and 2.10 and Table 2.3. The docked 3D structure of PfHDP–heme complex exhibited interaction of four residues, namely ASN35, LYS37, SER33, and THR88 of PfHDP with heme protein through four hydrogen bonds (H-bonds). The interacting H-bond distances were observed within the range of 2.723–3.943 Å. The docked 3D structure of PfDHFR–heme complex exhibited interaction of five residues, namely LYS69, LYS181, TYR183, TYR64, and TYR57 of PfDHFR with heme protein through five hydrogen bonds (H-bonds). The interacting H-bond distances were observed within the range of 2.065–4.107 Å.

4 Conclusion

Recent initiatives to develop more effective and affordable drugs and preventative vaccines have been launched with the goal of completely eradicating malaria. To this end, many pharmaceutical companies like Novartis and GlaxoSmithKline have reported over 20,000 "highly druggable" initial hits after screening chemical libraries of approximately 2 million small molecules for antimalarial properties. Furthermore, presently efforts in academia are centered on the elucidation of specific pathway

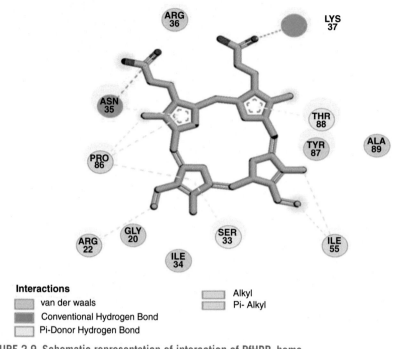

FIGURE 2.9 Schematic representation of interaction of PfHDP–heme.

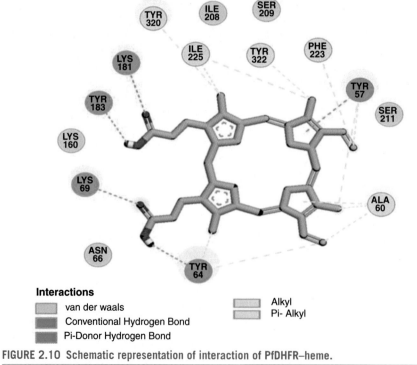

FIGURE 2.10 Schematic representation of interaction of PfDHFR–heme.

Table 2.3 Molecular interaction of PfHDP and PfDHFR with heme.

Parameters	PfHDP	PfDHFR
Binding energy (kcal/mol)	−7.47	−9.61
Bond type	Hydrogen bond	Hydrogen bond
Binding residues and atoms	ASN35–O	LYS69–O
	LYS37–O	LYS181–O
	SER33–π orbitals	TYR183–H
	ASN35–π orbitals	TYR64–H
	THR88–π orbitals	TYR57–π orbitals
Inhibitory constant (µM)	3.34	89.61
Intermolecular energy (kcal/mol)	−10.45	−12.6
Electrostatic energy (kcal/mol)	−3.43	−4.85

targets. One such high-throughput screening effort has been focused on hemozoin formation, which is a unique heme detoxification pathway found in the malaria parasite. This chapter discusses the importance of biosensor to diagnose several infectious as well as pathogenesis diseases like malaria.

The identification of diseased markers and predicting their effects by computational approaches has potential to generate personalized tools for the diagnosis and treatment of malaria. Previous studies have stated that both PfDHFR-TS and PfHDP are promising biomarker targets for malaria. Against this backdrop, in-depth analysis of molecular interaction of PfHDP and PfDHFR with heme was studied by using computational approach. The docking complexes showed binding energy to be −7.47 kcal/mol for PfHDP–heme and −9.61 kcal/mol for PfDHFR–heme. In-depth analysis of the docked complexes revealed the residues ASN35, LYS37, SER33, and THR88 of PfHDP to be involved in binding of ligand with the bond lengths varying from 2.723 to 3.943 Å. Similarly, the hydrogen bonding between PfDHFR and heme involved the residues LYS69, LYS181, TYR183, TYR64, and TYR57 and the bond lengths ranged from 2.065 to 4.107 Å. The binding energy was found to be −7.47 kcal/mol and −9.61 kcal/mol for docked PfHDP and heme and PfDHF-TS and heme complexes, respectively, signifying that heme could perform better sensing strength toward PfDHFR-TS as compared to PfHDP. It is believed that the current scenario of biosensor as discussed in this chapter along with the computational study will provide useful insights for design of therapeutic interventions for infectious diseases. In the future, novel methods must be developed to validate the mechanism of action of these hit compounds within the parasite.

References

[1] A. Kawamura, T. Miyata, Biosensors, in: Biomaterials Nanoarchitectonics, William Andrew Publishing, New York, 2016, pp. 157–176.

[2] W.R. Heineman, W.B. Jensen, Leland C. Clark Jr. (1918–2005), Biosens. Bioelectron. 8 (21) (2006) 1403–1404.

[3] C. Karunakaran, R. Rajkumar, K. Bhargava, Introduction to biosensors, in: Biosensors and Bioelectronics, Elsevier, Amsterdam, 2015, pp. 1–68.

[4] D.W.G. Morrison, M.R. Dokmeci, U. Demirci, A. Khademhosseini, Clinical applications of micro- and nanoscale biosensors, Biomed. Nanostruct. 1 (2008) 433–458.

[5] M.L.Y. Sin, K.E. Mach, P.K. Wong, J.C. Liao, Advances and challenges in biosensor-based diagnosis of infectious diseases, Expert Rev. Mol. Diagn. 14 (2) (2014) 225–244.

[6] R.F.O. França, C.C. Da Silva, S.O. De Paula, Recent advances in molecular medicine techniques for the diagnosis, prevention, and control of infectious diseases, Eur. J. Clin. Microbiol. Infect. Dis. 32 (6) (2013) 723–728.

[7] World Health OrganizationWorld Malaria Report 2019, World Health Organization, Geneva, (2019).

[8] N. Bhalla, P. Jolly, N. Formisano, E. Pedro, Introduction to biosensors, Essays Biochem. 60 (1) (2016) 1–8.

[9] K.V. Ragavan, K. Sanni, S. Swaraj, S. Neethirajan, Advances in biosensors and optical assays for diagnosis and detection of malaria, Biosens. Bioelectron. 105 (2018) 188–210.

[10] W. Jeon, L. Seonghwan, D.H. Manjunatha, B. Changill, A colorimetric aptasensor for the diagnosis of malaria based on cationic polymers and gold nanoparticles, Anal. Biochem. 439 (1) (2013) 11–16.

[11] A.K. Panigrahi, V. Singh, S.G. Singh, A multi-walled carbon nanotube–zinc oxide nanofiber based flexible chemiresistive biosensor for malaria biomarker detection, Analyst 142 (12) (2017) 2128–2135.

[12] W. Ittarat, S. Chomean, C. Sanchomphu, N. Wangmaung, C. Promptmas, W. Ngrenngarmlert, Biosensor as a molecular malaria differential diagnosis, Clin. Chim. Acta 419 (2013) 47–51.

[13] S. Lee, K.M. Song, W. Jeon, H. Jo, Y.B. Shim, C. Ban, A highly sensitive aptasensor towards Plasmodium lactate dehydrogenase for the diagnosis of malaria, Biosens. Bioelectron. 35 (1) (2012) 291–296.

[14] A. Hemben, J. Ashley, I. Tothill, Development of an immunosensor for PfHRP 2 as a biomarker for malaria detection, Biosensors 7 (3) (2017) 28.

[15] S. Lee, D.H. Manjunatha, W. Jeon, C. Ban, Cationic surfactant-based colorimetric detection of Plasmodium lactate dehydrogenase, a biomarker for malaria, using the specific DNA aptamer, Plos One 9 (7) (2014) e100847.

[16] K.B. Paul, S. Kumar, S. Tripathy, S.R.K. Vanjari, V. Singh, S.G. Singh, A highly sensitive self assembled monolayer modified copper doped zinc oxide nanofiber interface for detection of Plasmodium falciparum histidine-rich protein-2: targeted towards rapid, early diagnosis of malaria, Biosens. Bioelectron. 80 (2016) 39–46.

[17] M. de Souza Castilho, T. Laube, H. Yamanaka, S. Alegret, M.I. Pividori, Magneto immunoassays for Plasmodium falciparum histidine-rich protein 2 related to malaria based on magnetic nanoparticles, Anal. Chem. 83 (14) (2011) 5570–5577.

[18] N.W. Lucchi, A. Demas, J. Narayanan, D. Sumari, A. Kabanywanyi, S.P. Kachur, J.W. Barnwell, V. Udhayakumar, Real-time fluorescence loop mediated isothermal amplification for the diagnosis of malaria, PloS One 5 (10) (2010) e13733.

[19] E.V. Korotkaya, Biosensors: design, classification, and applications in the food industry, Foods Raw Mater. 2 (2) (2014) 161–171.

[20] S.J. Updike, G.P. Hicks, The enzyme electrode, Nature 214 (1967) 986–988, doi: 10.1038/214986a0.

[21] J. Wang, DNA biosensors based on peptide nucleic acid (PNA) recognition layers. A review, Biosens. Bioelectron. 13 (7–8) (1998) 757–762.

[22] P. Mehrotra, Biosensors and their applications—a review, J. Oral Biol. Craniofac. Res. 6 (2) (2016) 153–159.

[23] C. Divies, Remarques sur l'oxydation de l'éthanol par une electrode microbienne d'acetobacter zylinum, Ann. Microbiol. A 126 (1975) 175–186.

[24] W. Lorenz, K.D. Schulze, Application of transform-impedance spectrometry, J. Electroanal. Chem. 65 (1) (1975) 141–153.

[25] V. Scognamiglio, F. Arduini, G. Palleschi, G. Rea, Biosensing technology for sustainable food safety, TrAC Trends Anal. Chem. 62 (2014) 1–10.

[26] A.M. Foudeh, T.F. Didar, T. Veres, M. Tabrizian, Microfluidic designs and techniques using lab-on-a-chip devices for pathogen detection for point-of-care diagnostics, Lab Chip 12 (18) (2012) 3249–3266.

[27] B. Pejcic, R. De Marco, G. Parkinson, The role of biosensors in the detection of emerging infectious diseases, Analyst 131 (10) (2006) 1079–1090.

[28] D. Mabey, R.W. Peeling, A. Ustianowski, M.D. Perkins, Tropical infectious diseases: diagnostics for the developing world, Nat. Rev. Microbiol. 2 (3) (2004) 231.

[29] World Health Organization, The world health report 2004. Available from: http://www.who.int/whr/2004/en/report04_en.pdf, 2010.

[30] D. Mabey, R.W. Peeling, A. Ustianowski, M.D. Perkins, Diagnostics for the developing world, Nat. Rev. Microbiol. 2 (2004) 231–240.

[31] F.S. Rodrigues Ribeiro Teles, L.A. Pires de Tavora Tavira, L.J. Pina da Fonseca, Biosensors as rapid diagnostic tests for tropical diseases, Crit. Rev. Clin. Lab. Sci. 47 (3) (2010) 139–169.

[32] M. Strömberg, J. Göransson, K. Gunnarsson, M. Nilsson, P. Svedlindh, M. Strømme, Sensitive molecular diagnostics using volume-amplified magnetic nanobeads, Nano Lett. 8 (3) (2008) 816–821.

[33] P.D. Patel, (Bio)sensors for measurement of analytes implicated in food safety: a review, TrAC Trends Anal. Chem. 21 (2) (2002) 96–115.

[34] M. Nayak, A. Kotian, S. Marathe, D. Chakravortty, Detection of microorganisms using biosensors—a smarter way towards detection techniques, Biosens. Bioelectron. 25 (4) (2009) 661–667.

[35] C.K. Murray, J.W. Bennett, Rapid diagnosis of malaria, Interdiscip. Perspect. Infect. Dis. 2009 (2009) 1–7.

[36] M. Hawkes, K.C. Kain, Advances in malaria diagnosis, Expert Rev. Anti Infect. Ther. 5 (3) (2007) 485–495.

[37] A. Moody, Rapid diagnostic tests for malaria parasites, Clin. Microbiol. Rev. 15 (1) (2002) 66–78.

[38] C. Wongsrichanalai, M.J. Barcus, S. Muth, A. Sutamihardja, W.H. Wernsdorfer, A review of malaria diagnostic tools: microscopy and rapid diagnostic test (RDT), Am. J. Trop. Med. Hyg. 77 (Suppl. 6) (2007) 119–127.

[39] J.R. Clarke, Molecular diagnosis of HIV, Expert Rev. Mol. Diagn. 2 (3) (2002) 233–239.

[40] T. Bourlet, M. Memmi, H. Saoudin, B. Pozzetto, Molecular HIV screening, Expert Rev. Mol. Diagn. 13 (7) (2013) 693–705.

[41] B.M. Branson, State of the art for diagnosis of HIV infection, Clin. Infect. Dis. 45 (Suppl. 4) (2007) S221–S225.

[42] J.M. Katz, K. Hancock, X. Xu, Serologic assays for influenza surveillance, diagnosis and vaccine evaluation, Expert Rev. Anti Infect. Ther. 9 (6) (2011) 669–683.

[43] M.A. DiMaio, M.K. Sahoo, J. Waggoner, B.A. Pinsky, Comparison of Xpert Flu rapid nucleic acid testing with rapid antigen testing for the diagnosis of influenza A and B, J. Virol. Methods 186 (1–2) (2012) 137–140.

[44] N. Chauhan, J. Narang, S. Pundir, S. Singh, C.S. Pundir, Laboratory diagnosis of swine flu: a review, Artif. Cells Nanomed. Biotechnol. 41 (3) (2013) 189–195.

[45] V. Wiwanitkit, Dengue fever: diagnosis and treatment, Expert Rev. Anti Infect. Ther. 8 (7) (2010) 841–845.

[46] K.F. Tang, E.E. Ooi, Diagnosis of dengue: an update, Expert Rev. Anti Infect. Ther. 10 (8) (2012) 895–907.

[47] F.A. Al-Zamel, Detection and diagnosis of Mycobacterium tuberculosis, Expert Rev. Anti Infect. Ther. 7 (9) (2009) 1099–1108.

[48] D. Helb, M. Jones, E. Story, C. Boehme, E. Wallace, K. Ho, J. Kop, et al. Rapid detection of Mycobacterium tuberculosis and rifampin resistance by use of on-demand, near-patient technology, J. Clin. Microbiol. 48 (1) (2010) 229–237.

[49] A. Van Rie, L. Page-Shipp, L. Scott, I. Sanne, W. Stevens, Xpert MTB/RIF for point-of-care diagnosis of TB in high-HIV burden, resource-limited countries: hype or hope?, Expert Rev. Mol. Diagn. 10 (7) (2010) 937–946.

[50] D. Jani, R. Nagarkatti, W. Beatty, R. Angel, C. Slebodnick, J. Andersen, S. Kumar, D. Rathore, HDP—a novel heme detoxification protein from the malaria parasite, PLoS Pathog. 4 (4) (2008) e1000053.

[51] V.A. Nagaraj, B. Sundaram, N.M. Varadarajan, P.A. Subramani, D.M. Kalappa, S.K. Ghosh, G. Padmanaban, Malaria parasite-synthesized heme is essential in the mosquito and liver stages and complements host heme in the blood stages of infection, PLoS Pathog. 9 (8) (2013) e1003522.

[52] M.A. Biamonte, J. Wanner, K.G. Le Roch, Recent advances in malaria drug discovery, Bioorg. Med. Chem. Lett. 23 (10) (2013) 2829–2843.

[53] J.E. Hyde, Drug-resistant malaria—an insight, FEBS J. 274 (18) (2007) 4688–4698.

[54] J.R. Abshire, C.J. Rowlands, S.M. Ganesan, P.T.C. So, J.C. Niles, Quantification of labile heme in live malaria parasites using a genetically encoded biosensor, Proc. Natl. Acad. Sci. USA 114 (11) (2017) E2068–E2076.

[55] M. Wiederstein and M.J. Sippl, ProSA-web: interactive web service for the recognition of errors in three-dimensional structures of proteins, Nucleic Acids Res. 35 (2007) W407–W410.

[56] X.Y. Meng, H.X. Zhang, M. Mezei, M. Cui, Molecular docking: a powerful approach for structure-based drug discovery, Curr. Comput. Aided Drug Des. 7 (2011) 146–157.

[57] Accelrys Software Inc., Discovery Studio Visualizer, Release 4.0, 2013.

[58] A. Sali and T.L. Blundell, Comparative protein modelling by satisfaction of spatial restraints, J Mol Biol. 234 (1993) 779–815. https://doi.org/ 10.1006/jmbi.1993.1626.

[59] J. Ko, H. Park, L. Heo, C. Seok, GalaxyWEB server for protein structure prediction and refinement, Nucleic Acids Res. 40 (2012) W294–W297.

[60] D. Cojoc, S. Finaurini, P. Livshits, E. Gur, A. Shapira, V. Mico, Z. Zalevsky, Toward fast malaria detection by secondary speckle sensing microscopy, Biomed. Opt. Express 3 (2012) 991–1005.

[61] J. Shah, O. Mark, H. Weltman, N. Barcelo, W. Lo, D. Wronska, S. Kakkilaya, A. Rao, S.T. Bhat, R. Sinha, S. Omar, P. O'bare, M. Moro, R.H. Gilman, N. Harris, Fluorescence in situ hybridization (FISH) assays for diagnosing malaria in endemic areas, PLoS One 10 (2015) e0136726.

[62] S. Sathpathi, A.K. Mohanty, P. Satpathi, S.K. Mishra, P.K. Behera, G. Patel, A.M. Dondorp, Comparing Leishman and Giemsa staining for the assessment of peripheral blood smear preparations in a malaria-endemic region in India, Malar. J. 13 (2014) 512.

[63] World Health Organization, World Malaria Report 2016, World Health Organization, Geneva, (2016).

[64] W.L. DeLano, The PyMOL Molecular Graphics System, Delano Scientific, San Carlos, CA, 2009.

[65] S.F. Altschul, W. Gish, W. Miller, E.W. Myer, D.J. Lipman, Basic local alignment search tool, J. Mol. Biol. 215 (1990) 403–410.

[66] H.M. Berman, J. Westbrook, Z. Feng, G. Gililand, T.N. Bhat, H. Weissig, I.N. Shindyalov, P.E. Bourne, The protein data bank, Nucleic Acid Res. 28 (2000) 235–242.

[67] J. Yang, Y. Zhang, Protein structure and function prediction using I-TASSER, Curr. Protoc. Bioinformatics 52 (2015) 5.8.1–5.815.

[68] R.A. Laskowiski, M.W. MacArthur, D.S. Moss, J.M. Thornton, PROCHECK: a program to check the stereo chemical quality of protein structures, J. Appl. Crystallogr. 26 (1993) 283–291.

[69] D. Eisenberg, R. Luthy, J.U. Bowie, Verify 3D: assessment of protein models with three dimensional profiles, Methods Enzymol. 277 (1997) 396–404.

[70] C. Colovos, T.O. Yeates, Verification of protein structure: patterns of non-bonded atomic interaction, Protein Sci. 2 (1993) 1511–1519.

[71] N. Dey, V. Bhateja, A.E. Hassanien, Medical Imaging in Clinical Applications, Springer International Publishing, Cham, (2016).

[72] C. Bhatt, N. Dey, A.S. Ashour (Eds.), Internet of Things and Big Data Technologies for Next Generation Healthcare, Springer, New York, 2017.

[73] N. Dey, A. Ashour (Eds.), Classification and Clustering in Biomedical Signal Processing, IGI Global, Hershey, PA, 2016.

[74] N.Dey A.S. Ashour S.Borra Classification in BioApps: Automation of Decision Making, vol. 26 Springer, Cham, (2017).

Biosensors: a better biomarker for diseases diagnosis

Maheswata Moharana[a], Subrat Kumar Pattanayak[b]

[a]*Hydro & Electrometallurgy Department, CSIR-Institute of Minerals and Materials Technology, Bhubaneswar, India;* [b]*Department of Chemistry, National Institute of Technology, Raipur, India*

1 Introduction

Biosensors are kind of bioelectronic device which is generally used for bioanalysis purpose. Simultaneously, it can also provide real-time, accurate, and reliable information about the analytes. It may be defined as a compress analytical device which incorporates both biological and biologically active sensing elements either integrated within or closely connected with a physicochemical transducer [1]. The major applications of bioelectronics include the principles of electronics to biology and medicine. If we consider human body then nose, tongue, ears, eyes, and fingers can be included as sensors. Basically, sensors can give either qualitative or quantitative analysis. These are divided into three types: (1) physical sensors which are able to measure physical quantities like distance mass and temperature, (2) chemical sensors which measure an analyte by chemical or physical responses, for example, litmus paper, pH indicator solution, and pH meter are common type of sensors used for the test of acids and alkalis, and (3) biosensors which are able to measure an analyte by biological sensing elements/bioelement. To sense all the physical variables, sensors need to convert into a universal and easily accessible signal: usually called a voltage. The signal voltage continuously changes with time. A component called a transducer is responsible for this conversion. The resulting voltage signal in this conversion is usually called as an analog signal. This analog voltage signal transferred to computer or microprocessor, which can able to recognize digital signal only. There are two basic operating principles of biosensors: (1) sensing and (2) biological recognition. Hence, biosensor can also be defined as a device which is associated with three major components connected in a series: (1) bioreceptor, (2) transducer, and (3) microelectronics. The usual aim of biosensors is to generate a digital electronic signal that is proportional to the chemical concentrations or to the set of chemicals. Bioreceptors generally consist of immobilized biocomponents that are able to detect the specific target analyte. These biocomponents are mainly antibodies, nucleic acids, enzymes, and cell. The transducer acts as a converter which converts the biochemical signals

Smart Biosensors in Medical Care. http://dx.doi.org/10.1016/B978-0-12-820781-9.00003-6

into the electrical signals. The electrical signal is amplified and sent to a microelectronics and data processor. A measurable signal is then produced such as a digital display [2]. Biosensors are classified into three main components: (1) a bioreceptor, (2) a transducer, and (3) a signal processing system [2]. Biosensor is such a specific kind of sensing tool often called as biological sensor, which is made up of a transducer and a biological element such as an antibody, an enzyme, and a nucleic acid. Biological elements have connections with the analyte and the biological response which converted into the electrical signal by the transducer. All the biosensors are having a biological component which acts as a sensor and the electronic component detects as well as transmits the signal. Especially, biological materials get immobilized and a connection is made between the immobilized biological material and with the transducer. The analyte attach with the biological material to form a bonded analyte. In the present chapter, the authors highlighted about the interaction of different proteins (PDB ID: 4MPF and 3D78) with halauxifen-methyl, mesosulfuron-methyl andibutyl benzene-1,2-dicarboxylate, and 1,2-dimethoxy-4-prop-2-enylbenzene by molecular docking method.

2 Literature review

2.1 Components and basic principle of biosensor

Molecules like different antibodies, engineering proteins were acting as receptors in biosensor. Generally, antibody made up of thousands of amino acids was well arranged in specified ordered manner. Different enzymes were used as bioreceptors in biosensors. These were sensitive and selective. Specific antibodies play important role in immune sensors [3]. Specific antigens were able to identify by immune sensors. With the help of target antigen, antibodies are capable to detect the analyte. DNA biosensors are used in the biological recognition process. Because of their wide spread applications, DNA probe with a transducer has attracted considerable attention. Genosensors, align with oligonucleotides, can be monitored complementary DNA sequences. Nucleic acid can be determined by the help of biosensors sequences obtained from humans, bacteria, and viruses. For sensing point of view, cell-based sensors employed a single cell. The signal coming from the cell in response to an analyte is amplified by cell signaling machinery [4]. Therefore, the sensitivity is high. The immobilization of the receptor onto the transducer surface is assumed vital issue for the development of cell-based biosensors [5]. The function of transducer provided the output quantity with a relationship to input quality. There are different types of transducers, for example, optical, piezoelectric, calorimetric, and electrochemical [6]. Amperometric, conductimetric, potentiometric, and impedimetric are considered in the part of electrochemical transducers. In electrochemical biosensors, the chemical reaction occurred between biomolecule and target analyte [7]. Amperometric biosensor can identify the biochemicals by their enzyme-catalyzed hydrolysis [8]. The activity of potential difference that generated in the ion-selective membrane was measured by potentiometric transducers, whereas the resistivity or conductivity of solution in a biochemical process was measured by impedimetric transducers. Wheatstone bridge is associated with impedance biosensors [9]. Surface plasmon resonance (SPR) [10], fluorescence, and holographic are involved as part of in the optical biosensor. Fluorescence-based biosensors identified the frequency of the electromagnetic radiation emission. The recalibration process does not require the thermal biosensor [11], which is widely used in cosmetic, food, and pharmaceutical industries [12–15]. These types of biosensors were used in environmental, food industries [16] and DNA, as well as protein detection [17].

Biosensor is ubiquitous and able to monitor the properties of living systems. Although there are different applications of biosensors that are reported earlier in different literatures, the present study focuses on the biomedical applications for diagnosis of different diseases. Usually, most of the biomarker testing were time-consuming as well as costly. To avoid all these, faster, cost-effective devices are much more needed for analyses process [18]. Biosensor is a kind of device used to detect a biological analyte; it may be environmental or biological. Information such as the concentrations of the analyte at what level is transduced into an electrical signal that can be amplified, displayed, and analyzed. Analytes may include proteins (antigen, antibody, and enzyme), nucleic acid, or other biological or metabolic components (e.g., glucose). Biosensors can also find application in healthcare sectors, for example, it can be used to monitor blood glucose levels in diabetics, detect pathogens, and diagnose cancers. Besides, these biosensors have some environmental applications that include the detection of harmful bacteria or pesticides in air, water, or during preparation of foods. Defense and military sectors have also a strong interest in the development of biosensors as counter bioterrorism device that can able to detect elements of chemical and biological warfare to avoid potential exposure and infections. Moreover, the vision of the future of biosensors includes chip-scale devices placed on the human body for monitoring vital signs, correcting abnormalities, or even signaling a call for help in emergency. It is made up of three components: a recognition element, a signal transducer, and a signal processor that displays the results. The molecular recognition component detects a "signal" from the environment in the form of an analyte, and the transducer then converts the biological signal into an electrical output [19]. In terms of cancer, the analyte being detected is tumor biomarkers. Biosensors can detect whether a tumor is present, whether it is benign or cancerous, and whether the treatment is effective in reducing and eliminating cancerous cells. Biosensors which detect multiple analytes may prove to be useful in cancer diagnosis and monitoring.

2.2 Impact of nanotechnology on biosensor

Nanotechnology is an emerging field having enormous impact on biosensors in terms of diagnosis, prognosis, and monitoring of cancer. The application of nanotechnology to biosensor development improves the chances of detecting cancer earlier. The use of nanomaterials as imaging agents allows for more sensitive and precise measurement of cancerous tissues. Some common examples of nanomaterial are liposomes, dendrimers, buckyballs, and carbon nanotubes (CNT) that are used to improve cancer imaging [20]. Because of the small size, nanoparticles allow greater surface to volume ratio which favors better detection, imaging, and prognosis methods, and improved drug delivery to tumor sites. Nanocantilevers, nanowires, and nanochannels are some common examples of structures that have been used for the detection of cancer-specific molecular events and improve signal transduction [21]. Zhang et al. developed a nanowire-based biosensor for the detection of micro-RNAs which are important regulators of gene expression and are associated with cancer development. The traditional method of detecting micro-RNA is northern blot analysis which is of low sensitivity, time-consuming, and of high cost. Hence, the development of sensitive, inexpensive, easy-to-use biosensor for detecting cancer-related micro-RNAs represents an advance in the use of micro-RNAs as cancer biomarker [22]. The use of single-walled carbon nanotubes (SWCNTs) has also enhanced the detection capabilities of electrochemical biosensors, with increased sensitivity to enzymatic reactions. Nanotechnology has also enabled advances in optical biosensor technology in the form of surface-enhanced Raman scattering (SERS). SERS can measure up to 20 biomarkers at a time without any interference [23].

2.3 Biosensor recognition elements

In the past decades, biosensors usually used naturally occurring recognition elements that were purified from biological or environmental systems. With the advances in technology and synthetic chemistry, many biosensor elements that are used nowadays are synthesized in the laboratory with improved stability and reproducibility of the biosensor function. Some common examples of recognition elements are receptor proteins, antigens, antibodies, enzymes, and nucleic acids. Cell surface receptors are targets for drug delivery and are useful for monitoring the effectiveness of cancer therapeutics. Antigen and antibody-based recognition elements are the most rapid detection system among others. It is critically beneficial as this type of elements is the inherent specificity of antigen–antibody interactions and the target molecule does not need any purification before recognition. Among different enzymes, allosteric enzymes show great potential as recognition elements. The regulatory subunit takes on the role of the recognition element and the catalytic site becomes the transducer. One of the most advanced sensors of this class is the glucose sensor which uses glucose oxidase as the recognition element. Glucose oxidase catalyzes the oxidation of glucose in the presence of oxygen to yield gluconolactone and hydrogen peroxide. In this case, an amperometric transducer will measure the rate of elimination of oxygen molecule or the rate of formation of hydrogen peroxide and convert this into a glucose reading [24]. Aptamers are oligonucleotides that are selected as the emerging biosensor recognition element because of high binding affinities for the targets. A combinatorial chemistry-based technology termed SELEX (systematic evolution of ligands by exponential enrichment) has been developed to generate nucleic-acid ligands from a library of DNA and RNA oligonucleotides. Biosensors built on this type of systems have proven to be useful in the discovery of new biomarkers for early cancer diagnosis. RNAzymes and DNAzymes are common examples of allosteric aptamers useful for monitoring diseases especially Alzheimer's disease and diabetes. According to National Cancer Institute (NCI), biomarker can be defined as "a biological molecule found in blood, other body fluids, or tissues that is a sign of normal or abnormal process or of a condition or disease." Biomarker may also be used to monitor how well the body responds to a treatment for a disease. Cancer biomarkers are the most significant tools because of the early detection, accurate pretreatment staging, determining the response toward chemotherapy treatment, and monitoring disease progression [19,25–27]. Cancer is considered as the second most common cause of mortality and morbidity in western countries as well as in other countries. This is also called multifactorial molecular disease which includes multistage development of tumor cells. It can be caused either by a range of genetic or by environmental factors, such as exposure to carcinogenic chemicals or radiations, or have a microbiological cause, including bacterial (e.g., stomach cancer) or viral (e.g., cervical cancer) infection [27]. Also this is the leading cause of death in the world. Till date there have been 200 types of cancers known according to the NCI [28]. Around 222,520 new lung cancer patients were identified in the United States in 2010, and approximately 87% were non-small cell carcinomas [29,30]. More than 75% of the lung cancer cases were diagnosed in late because remains no practical ways to screen a large number of peoples at risk. Among various cancer diseases, lung cancer is a major health problem nowadays. Every year around 1.3 million new lung cancer cases and about 1.2 million lung cancer deaths have been reported in Europe and North America [31–34]. From a survey it has been found that in Eastern Europe mortality rate due to lung cancer is highest in case of males while Northern Europe and America have highest mortality rates in case of females. Some new lung cancer cases have expected to be increasing in some of the developing countries like India and China in next 5 years [35]. Early diagnostic of lung cancer with suitable treatment significantly improves the 5-year survival rate [36]. Prostate cancer is another important health issue because it is the third most common cancer in men. Prostate-specific antigen (PSA) has been identified as a biomarker to screen prostate cancer patients [37]. It has been shown that PSA is the most reliable tumor marker to detect prostate cancer in the early stage and to monitor the recurrence of the disease after treatment [38]. Breast cancer is one of the leading causes of cancer death in women. Classical clincopathological features indicating patient prognosis including tumor size, histological subtype and grade, lymph mode metastases, and lymphovascular invasion are derived from histological analysis of primary breast cancer samples. Biomarkers such as estrogen receptor (ER) and human epidermal growth factor receptor 2 (HER2) have been established and are assessed routinely for the treatment of breast cancer [39]. Although there are many advances in therapeutics that have been made earlier, quick diagnosis and early prevention are very critical issue for the control of this disease status. Biomarkers are the commonly promising indicative of a particular disease process and the cancer biomarkers play an important role in clinical diagnoses and evaluation of treatments for patients. Most of the immunoassay methods have been developed for the detection of cancer biomarkers. Because of the high sensitivity and selectivity nature of biosensors we are able to diagnose and manage targeted diseases at early stages and also facilitate timely therapy decision. At present, so many screening methods, such as chest radiograph (CRG), computed tomography (CT), low-dose CT (LDCT), magnetic resonance imaging (MRI), and positron emission tomography (PET), have been reported by the researchers earlier. Because of some drawbacks of these techniques such as being expensive and having low sensitivity for identifying cancer cells at early stages, some advanced techniques have been proposed. Among them, 18F-fluorodeoxyglucose PET/CT was applied for oncological imaging but unable to produce accurate results [40,41]. Magnetic induction tomography (MIT) is another proposed technique which is having advantages of low cost and high sensitivity [42]. Apart from this, biopsy is another common way to identify lung cancer but this technique is also expensive and requires trained physicians [43]. To avoid all these problems for detection biomarker-based techniques attracted much attention for early diagnosis of the cancer diseases. There are various stages between the discovery of biomarkers and their clinical use in patients with cancer. Post-genomic technologies and bioinformatics tools help to identify novel markers with the integration of relevant and appropriate information, such as gene expression, mutations, single nucleotide polymorphisms, and cancer biology. Verification of these biomarkers is accomplished by liquid chromatography mass spectroscopy to identify the top priority of the markers and eliminate inconsistent ones. Selected biomarkers are needed to be screened and investigated before clinical validation [44]. L. Wang reported that genetic and proteomics-based biomarkers are two major types of biomarkers, identified through tumor cells, urine, sputum, blood, or other body fluids. Proteomic biomarkers including Annexin II, APOA1, carcinoembryonic antigen (CEA), carbohydrate antigen 125 (CA125), carbohydrate antigen 19-9 (CA19-9), cytokeratin fragment 21-1 (CYFRA21-1), neuron-specific enolase (NSE), retinol binding protein (RBP), vascular endothelial growth factor (VEGF), and many more have been reported for lung cancer detection. CEF is a common proteomic biomarker which distinguishes malignant tissues and benign tissues. Enzyme-linked immunosorbent assay (ELISA) is another conventional approach for biomarker detection. As it requires a labeling process, it hinders monitoring of the probe/target interaction rapidly. To avoid this challenge, many high-sensitive and label-free detection techniques such as field effect transistors (FETs), SPR, and quartz crystal microbalance (QCM) have been investigated. Among all these techniques, FETs are attracting more attentions because they are compact, inexpensive, and able to integrate many sensors on the same chip.

2.4 Optical and electrochemical-based biosensors

Optical-based biosensors have been developed for early diagnosis of lung cancer markers and the techniques have been improved by applying nanotechniques and surface chemistry. Existing optical biosensors can be categorized into fluorescence, interferometric, SPR, optrode-based fiber, evanescent wave fiber, and resonant mirror optical biosensor. Most commercial platforms generally use fluorescence detection systems, while most of the research tools use grating coupler and resonant mirror systems [31]. SPR-based biosensors have been developed for biomolecular interactions. These sensors excite surface plasmon from the interface and measure the change in refractive index, which can be classified as label-free and real-time affinity reaction detection systems. Most of the SPR and FET-based biosensors have been developed for assaying CYFRA21-1 protein [45,46]. Wang et al. developed an optical system with high precision based on magnetic ELISA to detect CYFRA21-1. This system is a powerful tool for the rapid detection of lung cancer marker with advantages of compactness and high sensitivity [47]. Ribaut et al. also developed an innovative plasmonic optical fiber immunosensor which detects lung cancer marker cytokeratin 17. Their research outcomes offered significant contributions toward diagnosis of biomarkers in tissues in clinical environment [48]. Optical biosensors measure the variations in wavelengths of light. Change of the wavelengths in response to their cognition of the analyte is converted by optical transducers and electrical readings/ provides digital [49]. A new type of sensors with the use of an optical transducer is a biosensor which has photonic crystals. Such sensors are designated for capturing the very tiny volumes or light areas, which allow measurements at a high susceptibility, and to display results the light is transmitted to a higher electromagnetic field. This technique detects where and when the molecules or cells dissociate or bind to the surface made of crystal through the measurement of the light reflected in the crystal. Lase-induced fluorescence is another optical biosensor used for monitoring and diagnosis of throat cancer. The patient swallows the biosensor, laser beam is directed by the device, and on the surface of the esophagus a specific wavelength of light is emitted. The walls of the esophagus reflect light of specific wavelengths and the variation in the visualization of various wavelengths is influenced by this sensor and has found to be highly useful as compared to the conventional methods. Surgical biopsies and pain-associated recovery are prevented by these biosensors [19]. During the chemical reactions between the bioreceptors and analytes, there will be generations of ions or electrons. Thus, the detection of electrochemical biosensor is related to the measurable electrical properties of the solution [50]. Electrochemical biosensors can be classified into amperometric, potentiometric, voltammetric, conductometric, and FET. Recently so many electrochemical biosensors have been developed as lung cancer marker detection. These sensors normally contain semiconductors and screen-printed electrodes that detect various molecules including protein, antibody, DNA, antigen, and heavy metal ions. These biosensors are highly sensitive especially diagnosis of lung cancer markers. Besides this, electrochemical nanobiosensors offer a promising tool for diagnosis of molecules with some advantages, including low cost, more accurate, fast response, and high sensitivity. Altintas et al. developed a magnetic particle-modified capacitive sensor to detect some cancer markers such as CEA, CA15-3, and hEGFR. According to their experimental validation, it was found that CEA and hEGFR were detected at a concentration range of 5 pg/mL–1 ng/mL, while CA15-3 was detected in the range of 1–200 U/mL with high specificity which proposes that the sensor may become a useful tool for early detection of cancer [51]. Tabrizi et al. [52] also developed another high sensitivity electrochemical aptasensor based on carbon-gold nanocomposite modified screen-printed electrode. VEGF165 was observed in lung cancer patients by using this proposed sensor, which might have potential to become a powerful tool for lung cancer detection. Zamay et al. [53] also developed new aptamer-based electrochemical sensor to identify lung cancer. Mathur et al. [54] developed a biosensor for the early detection of cancer biomarker (HAPLN1) that expresses in malignant pleural mesotheliomas using surface-imprinted electrochemical method. They have used gold electrode in this biosensor that adsorbs target molecules by their thiol groups, recognizes the HAPLN1 biomarker at even picomolar concentrations, and has a response time of 2–5 minutes. Biomedical imaging process [55] and artificial neural networks [56] played important role in early detection of disease area. Development of algorithm and approaches based on computer programming also helped in biomedical sensing [57]. The Internet of Things [58], medical cyber-physical system [59] which associated different networks, exchanged data in healthcare area.

3 Molecular docking study

3.1 Results and discussion

The molecular docking study was performed by using AutoDock Vina software [60] where Lamarckian algorithm was used. The docking study [61–67] was performed with the flexible ligand and the rigid receptor. The visualization and image preparation were performed by using Discovery Studio Visualizer software [68]. Molecular interaction was studied of 4MPF [69] with halauxifen-methyl (PubChem ID: 16656802) and mesosulfuron-methyl (PubChem ID: 11409499). We also studied 3D78 with dibutyl benzene-1,2-dicarboxylate (Compound CID: 3026) and 1,2-dimethoxy-4-prop-2-enylbenzene (Compound CID:7127). All the molecular interaction was studied by docking method. The binding energy was found to be −3.7 kcal/mol for 4MPF-halauxifen-methyl. The molecular interaction between 4MPF-halauxifen-methyl would be obtained from intermolecular hydrogen bonding involved with binding residues GLU29, ASP8, GLN12, ARG15, ILE26, ALA21, ARG31, and LEU5. Residues like LYS77, ASP59, ARG107, GLU172, LYS131, PRO244, LYS149, and GLU97 involved in electrostatic interaction, whereas the bond lengths vary from 2.223 to 3.23 Å. THR227, PRO228, and TYR233 residues were involved in hydrogen bonding in 4MPF-mesosulfuron-methyl system. The binding energy was found to be −5.0 kcal/mol. It was clearly observed that mesosulfuron-methyl was showing more preferential interaction toward 4MPF compared to halauxifen-methyl. Interestingly, there was no electrostatic interaction found in the 4MPF-mesosulfuron-methyl. The active sites interactions of 4MPF with halauxifen-methyl and mesosulfuron-methyl were depicted in Figs. 3.1 and 3.2, respectively.

LYS 36, ASP33, HIS24, GLU23, LUS100, GLU62 residues were involved in electrostatic interaction in the 3D78 with dibutyl benzene-1,2-dicarboxylate. The bond lengths were varying from 1.55 to 5.56 Å. The binding energy was found−7.0 kcal/mol.

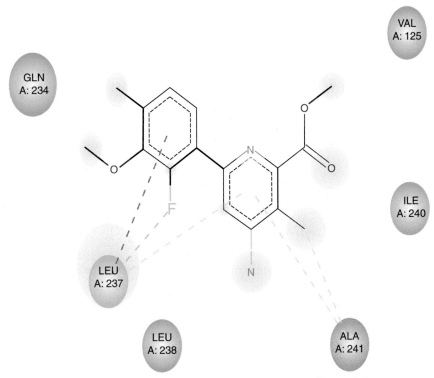

FIGURE 3.1 Two-dimensional representation between 4MPF-halauxifen-methyl.

Pink dotted lines are π-alky, *violet* represents π-sigma, and *blue* represents halogen bonds. The visualization and image preparation was performed by using Discovery Studio Visualizer software [68].

Residues like TRP4, ASP114, VAL5, PRO2, PHE10, PRO7, ASP11, GLU8, LEU12, GLU8, LEU12, and VAL9 were showing hydrogen bond. The bond lengths are varying from 1.90 to 3.29 Å. Two-dimensional representation of 3D78 [70] with dibutyl benzene-1,2-dicarboxylate was depicted in Fig. 3.3.

LYS17–ILE119, ARG19:HH11–GLU23:OE2, ARG19:HH21–ASP16:OD1, ARG81:HH12–ASP114:OD2, and ARG81:HH22–ASP114:OD1 were involved in salt bridge. ARG19:NH2–GLU54:OE1, HIS24:NE2–GLU23:OE1, LYS36:NZ–ASP33:OD1, and LYS100:NZ–GLU62:OE1 were associated in electrostatic interactions. The bond length was varying from 4.10481 to 5.56126 Å. TRP4:HE1–ASP114:O, VAL5:HN–PRO2:O, VAL9:HN–PRO6:O, PHE10:HN–PRO7:O, ASP11:HN–GLU8:O, LEU12:N–VAL9:O, VAL13:HN–VAL9:O, and ALA14:HN–ASP11:O were involved in conventional hydrogen bonding. The bond length was from 2.22 to 3.29 Å. The binding energy was found to be −6.1 kcal/mol. Two-dimensional interaction representation of 3D78 with 1,2-dimethoxy-4-prop-2-enylbenzene was depicted in Fig. 3.4. The negative value has shown better binding interaction.

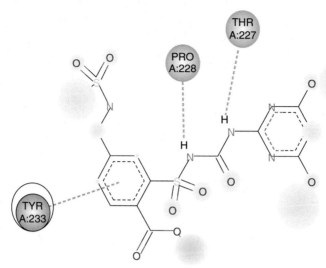

FIGURE 3.2 Two-dimensional representation of 4MPF-mesosulfuron-methyl ligand.

Pink, *violet*, and *blue dotted lines* represent π-alky, π-sigma, and hydrogen bonds, respectively. The visualization and image preparation was performed by using Discovery Studio Visualizer software [68].

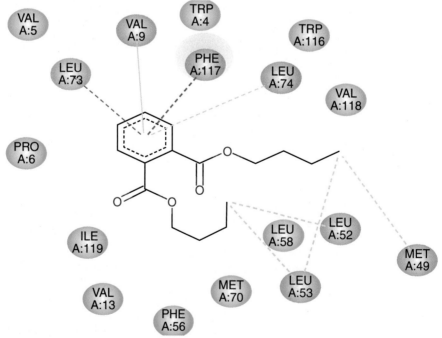

FIGURE 3.3 Two-dimensional representation of 3D78 with dibutyl benzene-1,2-dicarboxylate.

Pink dotted lines are π-alky, *violet* represents π-sigma, and *blue* represents halogen bonds. The visualization and image preparation was performed by using Discovery Studio Visualizer software [68].

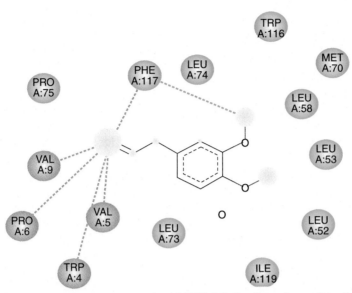

FIGURE 3.4 Two-dimensional representation of 3D78-1,2-dimethoxy-4-prop-2-enylbenzene.
Pink dotted lines are π-alky, *violet* represents π-sigma, and *blue* represents halogen bonds. The visualization and image preparation was performed by using Discovery Studio Visualizer software [68].

However, between dibutyl benzene-1,2-dicarboxylate and 1,2-dimethoxy-4-prop-2-enylbenzene the former was showing best binding interaction toward 3D78. The binding energy specified that dibutyl benzene-1,2-dicarboxylate could perform better sensing strength toward 3D78. The details binding energy and docking analysis results are given in Table 3.1.

Table 3.1 Molecular interaction of receptors and ligands.

Name of the PDB ID—ligand/ molecule	Binding energy (kcal/mol)	Residue involved in hydrogen bonding	Residue involved in electrostatic interaction
4MPF-halauxifen-methyl	−3.7	GLU29, ASP8, GLN12, ARG15, ILE26, ALA21, ARG31, LEU5	LYS77, ASP59, ARG107, GLU172, LYS131, PRO244, LYS149, GLU97
4MPF-mesosulfuron-methyl	−5.0	THR227, PRO228, TYR233	—
3D78-dibutyl benzene-1,2-dicarboxylate	−7.0	LYS17, ILE119, ARG19, GLU23, ASP16	LYS36, ASP33, HIS24, GLU23, LYS100, GLU62
3D78-1,2-dimethoxy-4-prop-2-enylbenzene	−6.1	LYS17, ILE119, ARG19, GLU23, ARG81, ASP114	HIS24, GLU23, LYS36, ASP33, LYS36, ASP33

4 Conclusion

Biosensors are having a biological component which acts as a sensor and the electronic component detects as well as transmits the signal. Biosensor is ubiquitous and able to monitor the properties of living systems. The main application of optical biosensor is to detect biological analytes and analysis of biomolecular interactions. These sensors normally contain semiconductors and screen-printed electrodes that detect various molecules including protein, antibody, DNA, antigen, and heavy metal ions. Molecular interaction was studied of 4MPF with halauxifen-methyl and mesosulfuron-methyl. We also studied 3D78 with dibutyl benzene-1,2-dicarboxylate and 1,2-dimethoxy-4-prop-2-enylbenzene. All the molecular interaction was studied by docking method. It was clearly observed that mesosulfuron-methyl was showing more preferential interaction toward 4MPF compared to halauxifen-methyl. However, between dibutyl benzene-1,2-dicarboxylate and 1,2-dimethoxy-4-prop-2-enylbenzene the former was showing best binding interaction toward 3D78. The binding energy specified that dibutyl benzene-1,2-dicarboxylate could perform better sensing strength toward 3D78.

References

[1] C. Karunakaran, R. Rajkumar, K. Bhargava, Introduction to biosensors, in: Biosensors and Bioelectronics, Elsevier, Amsterdam, 2015, pp. 1–68.

[2] V. Perumal, U. Hashim, Advances in biosensors: principle, architecture and applications, J. Appl. Biomed. 12 (2014) 1–15.

[3] R. Monošík, M. Stred'anský, E. Šturdík, Biosensors—classification, characterization and new trends, Acta Chim. Slovaca 5 (2012) 109–120.

[4] A.S. Khalil, J.J. Collins, Synthetic biology: applications come of age, Nat. Rev. Genet. 11 (2010) 367.

[5] F. Lagarde, N. Jaffrezic-Renault, Cell-based electrochemical biosensors for water quality assessment, Anal. Bioanal. Chem. 400 (2011) 947.

[6] M. Mehrvar, M. Abdi, Recent developments, characteristics, and potential applications of electrochemical biosensors, Analyt. Sci. 20 (2004) 1113–1126.

[7] D.R. Thevenot, K. Toth, R.A. Durst, G.S. Wilson, Electrochemical biosensors: recommended definitions and classification, Pure Appl. Chem. 71 (1999) 2333–2348.

[8] A. Heller, Amperometric biosensors, Curr. Opin. Biotechnol. 7 (1996) 50–54.

[9] M. Pohanka, P. Skládal, Electrochemical biosensors—principles and applications, J. Appl. Biomed. 6 (2008) 57–64.

[10] J. Homola, Present and future of surface plasmon resonance biosensors, Anal. Bioanal. Chem. 377 (2003) 528–539.

[11] S.P. Mohanty, E. Kougianos, Biosensors: a tutorial review, IEEE Potentials 25 (2006) 35–40.

[12] M.L. Antonelli, C. Spadaro, R.F. Tornelli, A microcalorimetric sensor for food and cosmetic analyses: l-Malic acid determination, Talanta 74 (2008) 1450–1454.

[13] S.G. Bhand, S. Soundararajan, I. Surugiu-Wärnmark, J.S. Milea, E.S. Dey, M. Yakovleva, B. Danielsson, Fructose-selective calorimetric biosensor in flow injection analysis, Anal. Chim. Acta 668 (2010) 13–18.

[14] K. Ramanathan, B. Danielsson, Principles and applications of thermal biosensors, Biosens. Bioelectron. 16 (2001) 417–423.

[15] S. Vermeir, B.M. Nicolai, P. Verboven, P. Van Gerwen, B. Baeten, L. Hoflack, V. Vulsteke, J. Lammertyn, Microplate differential calorimetric biosensor for ascorbic acid analysis in food and pharmaceuticals, Anal. Chem. 79 (2007) 6119–6127.

[16] S. Tombelli, M. Minunni, M. Mascini, Piezoelectric biosensors: strategies for coupling nucleic acids to piezoelectric devices, Methods 37 (2005) 48–56.

[17] M. Nirschl, A. Blüher, C. Erler, B. Katzschner, I. Vikholm-Lundin, S. Auer, J. Vörös, W. Pompe, M. Schreiter, M. Mertig, Film bulk acoustic resonators for DNA and protein detection and investigation of in vitro bacterial S-layer formation, Sens. Actuat. A Phys. 156 (2009) 180–184.

[18] M. Mascini, S. Tombelli, Biosensors for biomarkers in medical diagnostics, Biomarkers 13 (2008) 637–657.

[19] B. Bohunicky, S.A. Mousa, Biosensors: the new wave in cancer diagnosis, Nanotechnol. Sci. Appl. 4 (2011) 1–10.

[20] P. Grodzinski, M. Silver, L.K. Molnar, Nanotechnology for cancer diagnostics: promises and challenges, Expert Rev. Mol. Diagn. 6 (2006) 307–318.

[21] H.N. Banerjee, M. Verma, Use of nanotechnology for the development of novel cancer biomarkers, Expert Rev. Mol. Diagn. 6 (5) (2006) 679–683.

[22] G.J. Zhang, J.H. Chua, R.E. Chee, A. Agarwal, S.M. Wong, Label-free direct detection of MiRNAs with silicon nanowire biosensors, Biosens. Bioelectron. 24 (2009) 2504–2508.

[23] K.K. Jain, Applications of nanobiotechnology in clinical diagnostics, Clin. Chem. 53 (2007) 2002–2009.

[24] J.P. Chambers, B.P. Arulanandam, L.L. Matta, A. Weis, J.J. Valdes, Biosensor recognition elements, Curr. Issues Mol. Biol. 10 (2008) 1–12.

[25] C.F. Basil, Y. Zhao, K. Zavaglia, P. Jin, M.C. Panelli, S. Voiculescu, S. Mandruzzato, H.M. Lee, B. Seliger, R.S. Freedman, P.R. Taylor, Common cancer biomarkers, Cancer Res. 66 (2006) 2953–2961.

[26] American Society of Clinical Oncology, Clinical practice guidelines for the use of tumor markers in breast and colorectal cancer, J. Clin. Oncol. 14 (1996) 2843–2877.

[27] Z. Altintas, Molecular biosensors: promising new tools for early detection of cancer, Chest 8 (2015) 9.

[28] J. Li, S. Li, C.F. Yang, Electrochemical biosensors for cancer biomarker detection, Electroanalysis 24 (2012) 2213–2229.

[29] L. Zhang, H. Xiao, H. Zhou, S. Santiago, J.M. Lee, E.B. Garon, J. Yang, O. Brinkmann, X. Yan, D. Akin, D. Chia, Development of transcriptomic biomarker signature in human saliva to detect lung cancer, Cell. Mol. Life Sci. 69 (2012) 3341–3350.

[30] A. Jemal, R. Siegel, J. Xu, E. Ward, Cancer statistics, 2010, CA Cancer J. Clin. 60 (2010) 277–300.

[31] L. Wang, Screening and biosensor-based approaches for lung cancer detection, Sensors 17 (2017) 2420.

[32] W.Y. Lin, W.H. Hsu, K.H. Lin, S.J. Wang, Role of preoperative PET-CT in assessing mediastinal and hilar lymph node status in early stage lung cancer, J. Chin. Med. Assoc. 75 (2012) 203–208.

[33] Y. Zhang, D. Yang, L. Weng, L. Wang, Early lung cancer diagnosis by biosensors, Int. J. Mol. Sci. 14 (2013) 15479–15509.

[34] N. Hasan, R. Kumar, M.S. Kavuru, Lung cancer screening beyond low-dose computed tomography: the role of novel biomarkers, Lung 192 (2014) 639–648.

[35] V. Noronha, A. Ramaswamy, V.M. Patil, A. Joshi, A. Chougule, S. Kane, R. Kumar, A. Sahu, V. Doshi, L. Nayak, A. Mahajan, ALK positive lung cancer: clinical profile, practice and outcomes in a developing country, PLoS One 11 (2016) e0160752.

[36] T.A. Chiang, P.H. Chen, P.F. Wu, T.N. Wang, P.Y. Chang, A.M.S. Ko, M.S. Huang, Y.C. Ko, Important prognostic factors for the long-term survival of lung cancer subjects in Taiwan, BMC Cancer 8 (2008) 324.

[37] C. Stephan, M. Klaas, C. Müller, D. Schnorr, S.A. Loening, K. Jung, Interchangeability of measurements of total and free prostate-specific antigen in serum with 5 frequently used assay combinations: an update, Clin. Chem. 52 (2006) 59–64.

[38] A. Jemal, R. Siegel, E. Ward, T. Murray, J. Xu, C. Smigal, M.J. Thun, Cancer statistics, 2006, CA Cancer J. Clin. 56 (2006) 106–130.

[39] M.T. Weigel, M. Dowsett, Current and emerging biomarkers in breast cancer: prognosis and prediction, Endocr. Relat. Cancer 17 (2010) R245–R262.

[40] S. Chicklore, V. Goh, M. Siddique, A. Roy, P.K. Marsden, G.J. Cook, Quantifying tumour heterogeneity in 18 F-FDG PET/CT imaging by texture analysis, Eur. J. Nucl. Med. Mol. Imaging 40 (2013) 133–140.

[41] D. Ippolito, C. Capraro, L. Guerra, E. De Ponti, C. Messa, S. Sironi, Feasibility of perfusion CT technique integrated into conventional 18 FDG/PET-CT studies in lung cancer patients: clinical staging and functional information in a single study, Eur. J. Nucl. Med. Mol. Imaging 40 (2013) 156–165.

[42] H. Griffiths, W.R. Stewart, W. Gough, Magnetic induction tomography: a measuring system for biological tissues, Ann. NY Acad. Sci. 873 (1999) 335–345.

[43] B.Y. Cheng, Development of a chemiluminescent immunoassay for cancer antigen 15-3, Labeled Immunoass. Clin. Med. 23 (2016) 1348–1351.

[44] N.G. Frangogiannis, Biomarkers: hopes and challenges in the path from discovery to clinical practice, Transl. Res. 159 (2012) 197–204.

[45] H. Wang, X. Wang, J. Wang, W. Fu, C. Yao, A SPR biosensor based on signal amplification using antibody-QD conjugates for quantitative determination of multiple tumor markers, Sci. Rep. 6 (2016) 33140.

[46] S. Cheng, S. Hideshima, S. Kuroiwa, T. Nakanishi, T. Osaka, Label-free detection of tumor markers using field effect transistor (FET)-based biosensors for lung cancer diagnosis, Sens. Actuat. B Chem. 212 (2015) 329–334.

[47] B. Wang, J.T. Liu, J.P. Luo, M.X. Wang, Q.U. Shu-Xue, X.X. Cai, A three-channel high-precision optical detecting system for lung cancer marker CYFRA21-1, J. Optoelectron. Laser 24 (2013) 1849–1854.

[48] C. Ribaut, M. Loyez, J.C. Larrieu, S. Chevineau, P. Lambert, M. Remmelink, R. Wattiez, C. Caucheteur, Cancer biomarker sensing using packaged plasmonic optical fiber gratings: towards in vivo diagnosis, Biosens. Bioelectron. 92 (2017) 449–456.

[49] I.E. Tothill, Biosensors for cancer markers diagnosis, in: Seminars in Cell & Developmental Biology, Academic Press, New York, 2009.

[50] A. Gharatape, A. Yari Khosroushahi, Optical biomarker-based biosensors for cancer/infectious disease medical diagnoses, Appl. Immunohistochem. Mol. Morphol. 27 (2019) 278–286.

[51] Z. Altintas, S.S. Kallempudi, U. Sezerman, Y. Gurbuz, A novel magnetic particle-modified electrochemical sensor for immunosensor applications, Sens. Actuat. B Chem. 174 (2012) 187–194.

[52] M.A. Tabrizi, M. Shamsipur, L. Farzin, A high sensitive electrochemical aptasensor for the determination of VEGF165 in serum of lung cancer patient, Biosens. Bioelectron. 74 (2015) 764–769.

[53] G.S. Zamay, T.N. Zamay, V.A. Kolovskii, A.V. Shabanov, Y.E. Glazyrin, D.V. Veprintsev, A.V. Krat, S.S. Zamay, O.S. Kolovskaya, A. Gargaun, A.E. Sokolov, Electrochemical aptasensor for lung cancer-related protein detection in crude blood plasma samples, Sci. Rep. 6 (2016) 34350.

[54] A. Mathur, S. Blais, C.M. Goparaju, T. Neubert, H. Pass, K. Levon, Development of a biosensor for detection of pleural mesothelioma cancer biomarker using surface imprinting, PloS one 8 (2013) e57681.

[55] N. Dey, A.S. Ashour (Eds.), Classification and Clustering in Biomedical Signal Processing, IGI Global, Hershey, PA, 2016.

[56] N.DeyA.S.AshourS.BorraClassification in BioApps: Automation of Decision Makingvol. 26Springer, Cham, (2017).

[57] N. Dey, V. Bhateja, A.E. Hassanien, Medical Imaging in Clinical Applications, Springer International Publishing, Cham, (2016).

[58] N. Dey, A.S. Ashour, C. Bhatt, Internet of things driven connected healthcare, in: Internet of Things and Big Data Technologies for Next Generation Healthcare, Springer, Cham, 2017, pp. 3–12.

[59] N. Dey, A.S. Ashour, F. Shi, S.J. Fong, J.M.R. Tavares, Medical cyber-physical systems: a survey, J. Med. Syst. 42 (4) (2018) 74.

[60] X.Y. Meng, H.X. Zhang, M. Mezei, M. Cui, Molecular docking: a powerful approach for structure-based drug discovery, Curr. Comput. Aided Drug Des. 7 (2011) 146–157.

[61] A. Bissoyi, S.K. Pattanayak, A. Bit, A. Patel, A.K. Singh, S.S. Behera, D. Satpathy, Alphavirus nonstructural proteases and their inhibitors, in: P. Gupta (Ed.), Viral Proteases and their Inhibitors, Academic Press, USA, 2017, pp. 77–104.

[62] S.M. Hiremath, A. Suvitha, N.R. Patil, C.S. Hiremath, S.S. Khemalapure, S.K. Pattanayak, V.S. Negalurmath, K. Obelannava, Molecular structure, vibrational spectra, NMR, UV, NBO, NLO, HOMO-LUMO and molecular docking of 2-(4, 6-dimethyl-1-benzofuran-3-yl) acetic acid (2DBAA): experimental and theoretical approach, J. Mol. Struct. 1171 (2018) 362–374.

[63] S.M. Hiremath, A. Suvitha, N.R. Patil, C.S. Hiremath, S.S. Khemalapure, S.K. Pattanayak, V.S. Negalurmath, K. Obelannavar, S.J. Armaković, S. Armaković, Synthesis of 5-(5-methyl-benzofuran-3-ylmethyl)-3H-[1, 3, 4] oxadiazole-2-thione and investigation of its spectroscopic, reactivity, optoelectronic and drug likeness properties by combined computational and experimental approach, Spectrochim. Acta A Mol. Biomol. Spectrosc. 205 (2018) 95–110.

[64] A. Chand, P. Chettiyankandy, M. Moharana, S.N. Sahu, S.K. Pradhan, S.K. Pattanayak, S.P. Mahapatra, A. Bissoyi, A.K. Singh, S. Chowdhuri, Computational Methods for Developing Novel Antiaging Interventions, Springer, Singapore, 2018.

[65] S.N. Sahu, S.K. Pattanayak, Molecular docking and molecular dynamics simulation studies on PLCE1 encoded protein, J. Mol. Struct. 1198 (2019) 126936.

[66] S.N. Sahu, M. Moharana, R. Sahu, S.K. Pattanayak, Impact of mutation on podocin protein involved in type 2 nephrotic syndrome: insights into docking and molecular dynamics simulation study, J. Mol. Liq. 281 (2019) 549–562.

[67] J. Panda, J.K. Sahoo, P.K. Panda, S.N. Sahu, M. Samal, S.K. Pattanayak, R. Sahua, Adsorptive behavior of zeolitic imidazolate framework-8 towards anionic dye in aqueous

media: combined experimental and molecular docking study, J. Mol. Liq. 278 (2019) 536–545.

[68] DS Visualizer, Accelrys software inc., Discovery Studio Visualizer 2, 2005.

[69] M. Shahbaaz, S. Kanchi, M. Sabela, K. Bisetty, Structural basis of pesticide detection by enzymatic biosensing: a molecular docking and MD simulation study, J. Biomol. Struct. Dyn. 36 (2018) 1402–1416.

[70] K. Langeswaran, J. Jeyaraman, R. Mariadasse, S. Soorangkattan, Insights from the molecular modeling, docking analysis of illicit drugs and bomb compounds with honey bee odorant binding oroteins (OBPs), Bioinformation 14 (2018) 219–231.

Smart biosensors for an efficient *point of care* (PoC) health management

Sahar Qazi, Khalid Raza

Department of Computer Science, Jamia Millia Islamia, New Delhi, India

1 Introduction to biosensors

Biosensors are devices based on receptor–transducer property and can convert biological information into useful and important physical/chemical or electrical signals [1]. They are very selective and sensitive when compared to their contemporary detection and diagnostic approaches, thus have been widely appreciated by medical practitioners and patients as well. What we mean by *sensitivity* is that the biological component, for instance, tissue, enzymes, organelles, cell receptors, antibodies, nucleic acids, etc., is a biological entity which interacts sensitively with the subject of study, which in turn then generates a signal which is detected by the transducer [2]. With the use of such highly sensitive receptors and detectors, the value of every biosensor is quite expensive. Fortunately, with mega demand of commonly used biosensors, some of them can be purchased at an affordable price.

Biosensors are very common these days and have been seen as a *trend* for everyone—be it children, adults, or aged people. These devices are usually *wearable* making it easy for people to maintain their healthcare regimen periodically [2]. The common wearable device these days is the "FitBit" activity tracker band which monitors the number of steps trotted by the individual, heart rate, sleep quality, steps climbed, and other personal measures engaged with fitness [3]. Glucometers, like Accucheck or Dr. Morepen blood glucose monitors, is another sort of biosensors which are used to monitor the glucose level in body [4]. Today, with the rise of healthcare awareness, many industrial and IT companies have launched biosensors for almost every healthcare-related problem.

The commercial industry is on a boom these days as everyone desires for a homely treatment. Even the consultants these days suggest keeping commercially available and easy-to-use biosensor devices for periodic health checkups at home. Fig. 4.1 represents the commercially available and commonly used smart biosensors which have made healthcare management easy for everyone. Glucometers, blood-sugar monitors, blood pressure monitors, fertility monitors, saliva-based glucose level monitors, and digital bandages for vital signs such as heart rate, fitness bands,

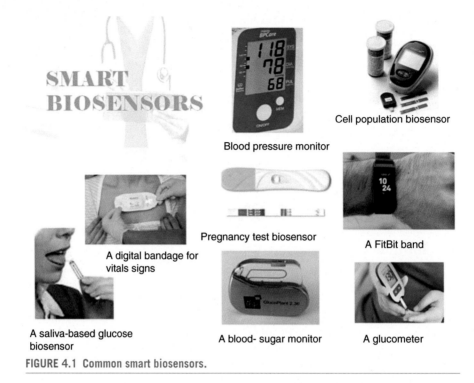

SMART
BIOSENSORS

Cell population biosensor

Blood pressure monitor

Pregnancy test biosensor

A FitBit band

A digital bandage for
vitals signs

A saliva-based glucose
biosensor

A blood- sugar monitor

A glucometer

FIGURE 4.1 Common smart biosensors.

etc., are some of the smart biosensors which make point of care (PoC) possible for anyone, anytime, and at anyplace [2]. The medical fraternity advertises the use of connected smart biosensors which is simply syncing the biosensor reading with one's mobile phones [5].

1.1 A biosensor system: receptor–transducer relationship

A biosensor system is composed of mainly three components: (1) a recognition region, (2) a transducer element, and (3) an electronic circuitry encapsulating an amplifier, a processor, and a visualizing display. Many times, the transducers and electric circuitry can be combined as one, for instance, in CMOS-based microsensor environments [6,7]. The receptor, as the name suggests, receives the biological analytes (biomolecules) of interest. This interaction (biological signal) can either be in the form of light, charge, mass change, or heat between the biological analytes and the receptor (bioreceptor) is analyzed by the transducer by throwing an output signal (physical/chemical signal) to the electronic circuitry or simply the signal processor producing an electrical signal. This electrical signal is then amplified and thereby visualized on the display monitor. The architecture of a biosensor is simple, and its basic desideratum is to rapidly test and detect where the biological analyte was procured. For further lucidity, we represent biosensor architecture in Fig. 4.2.

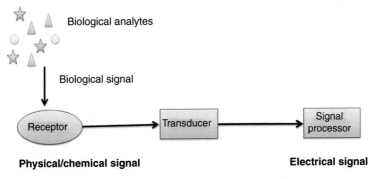

FIGURE 4.2 The architecture of a typical biosensor.

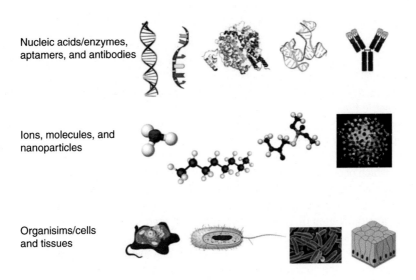

Nucleic acids/enzymes, aptamers, and antibodies

Ions, molecules, and nanoparticles

Organisims/cells and tissues

FIGURE 4.3 Common types of bioreceptors in biosensors.

1.2   Kinds of biological receptors

A bioreceptor is a biomolecule which can successfully identify the biological analyte. A bioreceptor can either be an enzyme, tissues or cells, any of the nucleic acids
RNA or DNA, antibodies, etc. The most commonly used bioreceptors in biosensors
are shown in Fig. 4.3. It is being reiterated here that bioreceptors are biological mac-
romolecules which attach biological samples/analytes onto their surface producing a
biological signal which is passed onto the biotransducer for further processing. Once
analytes get themselves latched on the surface of these bioreceptors, this interaction
can either be in the form of light, heat, pH, change in mass, change of charge, etc.
Henceforth, this interaction is termed as *biorecognition* [8].

1.3   Kinds of biological transducers

Transducers in a biosensor play a vital role in the entire conversion of one form of energy to another. One must keep Fig. 4.2 in mind while reading this section as it makes the entire hierarchical process a story leading from one signal conversion to another. Once the biorecognition event takes place, the physical/chemical signal from the bioreceptor gets passed onto the transducer which converts it into a measurable signal. This conversion of one form of energy signal to another by the biotransducer is termed as *signalization* [8]. Generally, biotransducers generate either optical or electrical signals which are somewhat proportional to the concentration of analyte–bioreceptor interaction. A biosensor's functioning is very much dependent on the type of biotransducer it has. There are mainly three kinds of biotransducers, namely electrochemical, ion channel, and reagentless fluorescent biotransducers. These have been described in much detail in the following subsections.

1.3.1   Electrochemical biotransducers

Electrochemical biosensors have a *biorecognition* component which actively interacts with the analyte of interest and generates an electrical signal, namely directly proportional to the analyte–bioreceptor amount. This signaling event can be rectified and eventually categorized as amperometric, potentiometric, impedance, and conductometric [9–11].

Amperometric transducers, as the name suggests, depend on the changes in electric current. They rectify changes in the current flow as a result of electrochemical redox (reduction/oxidation) reactions. This current flow change represents the interaction which occurs between the analyte and the bioreceptor that reflects the reaction occurring between the bioreceptor molecule and analyte restricted by the mass shift of analyte to the electrode. On the other hand, *potentiometric* transducers quantify charge agglomeration of an electrochemical cell, composed of an ion-selective electrode that interacts with charged ions. This affiliation causes agglomeration of charge potential to the reference electrode. The charge potential is directly proportional to the logarithm of the concentration of the analyte which is based on the Nernst equation. Impedance transducers engage with resistance and capacitive changes which usually take by place by biorecognition event. A three-tier-based electrode is specific enough for the biological analyte for immobilization on the surface of bioreceptor during the biorecognition process. To this, voltage is applied and then its current flow is measured. The impedance analyzer is used to measure the impedance change and controls the overall activity. Quite varying from the rest of its counterparts, *conductometric* transducers employ quantifying changes in the conductive properties of the reaction medium. The biorecognition changes the ionic concentration which changes the reaction media, in turn leading changes in the conductive flow [9–12].

1.3.2   Ion-channel switch biotransducers

Ion-channel switch biotransducers have taken inspiration from the ion channels which let the movement of molecules in and out easily through the membranes.

These biotransducers are very flexible and can even quantify picomolar concentrations of biomolecules. These biotransducers imitate the stimuli-based sensory responses. This biotransducer is not only applicable to ions, molecules, and nanoparticles but also can be employed for nucleotides and antibodies [13].

These transducers consist of an impedance unit which can be regulated according to the requirement and is a pivotal part of a microelectronic circuit. As mentioned earlier, these transducers are very flexible and can be used on blood samples as well. Generally, ion-channel switch transducers are applied for the identification of viruses, antibodies, DNA electrolytes, drugs, small molecules, etc. [14].

1.3.3 Reagentless fluorescent biotransducers

These biotransducers are developed and used mainly for detection of biologically important molecules. As the name suggests, these biosensors have fluorescent tags or simply *fluorophores*, which are nothing but protein entities, providing specificity by attaching the analyte of interest onto them. This bond between the protein entity (fluorophore) and the analyte is identified by the phenomenon of fluorescence which takes place during this interaction. Reagentless fluorescent biotransducers have been mainly used for two biochemical processes, namely GDP and Pyrophosphate. The main aim of such biotransducers is to identify and measure the release of analytes from cellular, biochemical, and metabolic reactions in real time along with discerning high sensitivity and specificity. These biotransducers are also helpful in providing high-resolution signals which are helpful in the analysis of the rate of the cellular/biochemical/metabolic reactions, nearer to the physiological environment [15].

2   Types of biosensors

Based on sensor devices and biological material used, biosensors are classified as shown in Fig. 4.4.

2.1   Electrochemical biosensors

These sensors are based on the reaction of enzymatic catalysis (such as redox enzymes) which either produces or consumes electrons. The electrochemical biosensors measure electric current, ionic, or condolence changes due to bioelectrodes. Electrochemical biosensors are further classified into four subtypes, namely *amperometric biosensors*, *potentiometric biosensors*, *impedimetric* biosensors, and *voltammetric biosensors*.

Amperometric biosensors: These subtypes of sensors are self-contained incorporated devices based on the movement of electrons, that is, electric current determination, as a result of the enzyme-catalyzed redox reaction. The "Clark oxygen" electrode that finds the reduction of O_2 is the simplest form of amperometric biosensors. In general, these biosensors have reaction times, sensitivities, and energetic ranges. Blood-glucose biosensors, widely used by diabetic patients, are a good example of

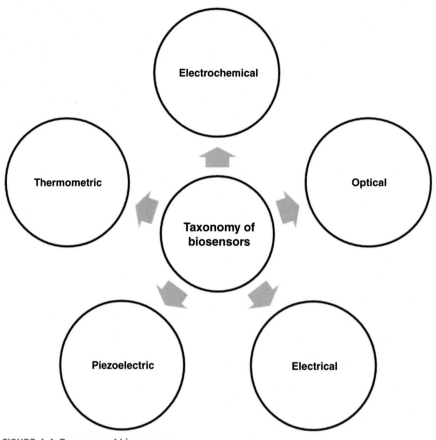

FIGURE 4.4 Taxonomy of biosensors.

amperometric biosensor. Furthermore, in order to assess the freshness of fishes, these biosensors have been developed which compare the accumulated ionosine and hypoxanthine with other nucleotides.

Potentiometric biosensors: In these biosensors, ion-selective electrodes are used to determine the changes in ionic concentration. The most widely used ion-selective electrode is pH electrode because several enzymatic reactions engross either release or absorption of hydrogen ions. Due to the low cost of ion-selective field-effect transistor (ISFET), these can be used to miniature potentiometric biosensors [16]. In fact, potentiometric assays are based on tracking potential/pH variations which are applied to the determination of several organic and inorganic compounds such as sugar, urea, pesticides, antibiotics, ammonia, CO_2, etc.

Impedimetric biosensors: Impedimetric biosensors are sensitive sensors because it uses Electrochemical Impedance Spectroscopy (EIS) [17]. It has been proved to be an effective technique for foodborne pathogenic bacteria detection and on-the-spot detection due to its rapidity, portability, and sensitivity, and observed an increasing application in the last decade [18]. These sensors are capable to monitor catalyzed reactions of enzymes, or biomolecular recognition of specific binding proteins, receptors, lectins, nucleic acids, whole cells antibodies, etc. [19]. Impedance detection approaches can be broadly divided into two classes based on the present or absent of specific biorecognition elements: (1) to measure the impedance change due to binding of targets to bioreceptors (such as antibodies and nucleic acids) immobilized onto the surface of the electrode, and (2) based on metabolites produced by bacterial cells. The impedimetric biosensors, its types, and architecture have been reviewed in Guan et al. [17], Wang et al. [18], and Bahadır and Sezgintürk [19].

Voltammetric biosensors: Such biosensors are based on *voltammetric method* which uses a carbon glue electrode customized with hemoglobin (Hb) in order to notice acrylamide. This type of electrode encounters reversible oxidation, or Hb reduction procedure. Voltammetric methods can be roughly classified into *cyclic voltammetry*, *differential pulse voltammetry*, *stripping voltammetry*, *linear sweep voltammetry*, etc. The most commonly used voltammetry methods are cyclic voltammetry and differential pulse voltammetry.

2.2 Optical biosensors

It uses optical transducers which are based on fluorescence or optical diffraction [20]. Due to selectivity and sensitivity properties, fluorescence is usually deployed for biosensing. A fluorescence-based optical sensor can detect the change in the frequency of electromagnetic radiation emission. On the other hand, optical diffraction-based sensors use a silicon wafer coated with a protein. Fiber-optics biosensor uses optical fibers for signal transduction. Surface plasma resonance (SPR) biosensors are also optical sensors which utilize specific electromagnetic waves to detect interactions between an analyte in solution. The application of SPR biosensor is to detect biological analytes and biomolecules interaction [21]. Also, an immuno-sensor based on localized surface plasma resonance (LSPR) on gold nanoparticle was used to detect casein in milk [22].

2.3 Electrical biosensors

Electrical biosensors are widely used toward miniaturized, sensitive, and portable detection protocol for biomarkers [23]. The electrical biosensors may be *conductometric* or *ion-sensitive*. In conductometric protocol, when ions or electrons are generated during biochemical reaction, the overall conductivity of the solution changes and its electrical conductance/resistance are measured. However, conductance measurements have relatively low sensitivity. On the other hand, ion-sensitive biosensors are based on ISFETs. The ISFET is a metal oxide semiconductor field-effect transistor which has an ion-sensitive surface. Due to interaction between ions and semiconductor, the surface electrical potential changes which can be measured [20,24].

2.4 Piezoelectric biosensors

It is based on the coupling of the bioelements with a piezoelectric unit, which is generally a quartz crystal coated with gold electrodes. Some of the materials that exhibit piezoelectric effects are quartz, lithium niobate, oriented zinc oxide, and aluminum nitride. It facilitates label-free detection of molecules. These sensors have several potential applications in food analysis, environmental, and analysis [20].

2.5 Thermometric biosensors

Thermometric biosensors are made by immobilization of biomolecules onto temperature sensors, where analyte comes in contact with biocomponent and reaction heat is measured. The temperature is measured by a thermistor, also known as *enzyme thermistor*. These types of sensors do not need frequent recalibration and are insensitive to the optical and electrochemical properties of samples. These sensors are used for food cosmetics and pharmaceutical analysis.

3 Applications of biosensors

The *smartness* of biosensors has already been described in the previous sections along with their types and basic principle. It's time to pay heed to their umbrella wide implications which are presently being cherished not only by the industrialists, academicians, and public but also by the environmentalists, physicians/medical consultants, etc. High specificity and sensitivity, reusability, affordability, easy usage, and dependence of basic parameters make a biosensor *smart*. Biosensors have been applied in myriad domains like environmental studies, biotechnology and food industry, healthcare, marine domain, etc. It has been observed that they give a much better and efficient environment when compared to traditional methods [24].

3.1 Detection of heavy materials

The use of biosensors for rectification and quantification of heavy metal ions has been one of the most crucial applications of biosensors. Heavy metal pollution is a very serious concern to human health as the metal ions do not degrade completely and get retained in the milieu. Traditional methods, for instance, cold vapor atomic absorption spectrometry and coupled plasma mass spectrometry, have been appreciated in the past but endured many limitations such as requirement of a trained individual, expensive infrastructure, and machinery. One of the crucial loopholes of these techniques was the fact that these were laboratory bound. Biosensors replaced these techniques by having an edge of specificity and sensitivity, affordable prices, easy to employ, mobility of the device, and rapid real-time signal processing. Smart biosensors having efficient compatibility with both proteins and whole-cell-based approaches can be utilized for the analysis of heavy metal ions [25].

3.2 Biotechnology, fermentation, and food industry

Everyone wants a good and healthy quality of lifestyle, which includes not only great living but also healthy food eating habits. Healthy foods are determined only by their quality of manufacturing and processing. Previously, chemical experimentations and spectroscopy were being employed in assessing the quality of the food, which was very expensive, time-taking, and tedious processes. Henceforth, alternative strategies which held the potential for efficient monitoring at a pocket-friendly price were yearned. Eventually, biosensors which were sensibly selective, cost-friendly, simple, and real-time based were developed and brought into the domains of food industry management [26]. This ideology grew huge and researchers developed enzymatic biosensors based on cobalt phthalocyanine for rapid aging of beer as it provided efficient management and storage of beer accordingly [27]. Moreover, many biosensors were developed for identification of microorganisms in food products, such as *Escherichia coli* is being used as a biomarker for decomposed disheveled contaminants in food [28], quantified by identifying alterations in pH caused by ammonia, namely generated by urease—*E. coli* antibody conjugate through a potentiometric biosensor. Biosensors can manage fermentation processes and can generate products at a rapid pace automatically. Currently, commercial biosensors can rectify crucial criteria such as glucose, lactate, lysine, etc. Ion-channel-based biosensors are also used to detect alterations in biochemical composition of products making eyeballs roll toward their simplicity and robustness for facilitating an efficient and regulated fermentation process [29].

The dairy industry employs more of enzymatic biosensors wherein the biosensor is based on an imprinted carbon electrode screen fitted into a flow cell. The enzymes get latched on the electrodes by a photo-cross-linkable polymer allowing the flow-dependent enzymatic biosensor to monitor the pesticides in milk [30]. Furthermore, electrochemical-based biosensors are being used to differentiate between natural and artificial sweeteners [24].

3.3 Microbial detection in environment

Environment specialists are on tiptoes because of humongous pollution. Our environment is full of microbial pollutants which require a stringent purification as they create havoc in the milieu, eventually disturbing the entire ecosystem. Whole-cell/organism-based biosensors are being widely used today because of their user and pocket-friendly characteristics, and can be seen as a vital option for the detection of microbes which can maneuver in order to acquire better stability under atrocious environments [31]. Generally, electrochemical biosensors are used to offer high sensitivity and rapid microbial detection in environments. Electrochemical and optical biosensors employ live microorganisms as their basic mechanism. Electroactive species which are either consumed or produced by microbes are detected by conductimetric, amperometric, impedimetric, or potentiometric biosensors, respectively, while optical biosensors are used for the quantification of fluorescent, chemiluminescent, or other optical products produced from microbes [32].

3.4 Ozone biosensing

Everyone is aware of the fact that ozone shields harmful ultraviolet (UV) rays. The harmful UV rays fall on the Earth's surface. Researchers are more interested in finding out how much UV rays penetrate the Earth's surface. One of the particular concerns is the question of how deeply they penetrate into the seawater and how it affects marine organisms, floating microorganism (plankton), and the marine ecosystem. A researcher from the Laboratory of Radiobiology and Environmental Health (University of California, San Francisco) worked in the Antarctic Ocean wherein she submerged plastic cases full of *E. coli* strains. The bacterial species cannot repair its damaged DNA, henceforth; it was a clever strategy to use them as a *live* biosensor. This strategy discerned genetic damages caused by the UV rays, thus providing an approach for quantifying UV ray penetration and intensity.

3.5 Point-of-care (PoC) monitoring

3.5.1 Blood-glucose-level measurement

It has been prophesized that by 2050, around 48.3 million people in the United States will endure diabetes mellitus [33], which is the most common endocrine disorder of carbohydrate metabolism leading to severe health problems for most developed societies. It also has been observed that diabetes can also cause mortality in the majority of people. The basic reason for the growing number of individuals to diabetes is the sedentary lifestyle that we carry today. In order to monitor glucose level, commercial industries have introduced *glucometers* (blood-glucose monitors) which help people to maintain their eating habits and health-related issues [34]. Self-regulation/monitoring of blood glucose (SRBG/SMBG) is being employed as a pivotal approach for the management of diabetes [35–40]. The goal of this approach is to help the patient to attain and regulate normal blood-glucose concentrations so as to curb other related disorders such as retinopathy, nephropathy and neuropathy, stroke and coronary artery disease, etc. [34].

3.5.2 Interferometric reflectance imaging (IRIS)

Interferometric reflectance imaging sensor (IRIS) is one of the widely employed biosensor systems which hold the capacity of high-throughput multiplexing of protein–protein, protein–DNA, and DNA–DNA interactions without employing fluorescent tags. Biological probes are spotted onto the sensing surface which is made functional using Si/SiO_2 substrates. Quantification of bimolecular mass which has been accumulated can be easily measured using IRIS biosensor system [41,42]. Due to its flexibility, it has been applied for many diagnostic and forensic researchers such as SNP identification, immunoassays, basic biomolecular interaction analyses, biomolecular kinetics studies, etc. [42].

3.5.3 DNA biosensors for forensic/biomedical and agricultural research

Till now we have covered biosensors which were either electrochemical, optical, fluorescent tags based, etc., but, in this section, we describe how DNA can be used as a biosensor. DNA is one of the nucleic acids, is double-stranded, and is ubiquitous in all except for some retroviruses [43]. Since it makes the genetic composition of living beings, it has the potential to act as a molecular recognizer as it binds to single-stranded complementary strands; this property of DNA gets exploited when making a DNA biosensor [44,45]. DNA biosensors are mainly based on small probes such as single-stranded oligonucleotides, aptamers, short peptides, and a few DNA-relevant protein molecules [46]. DNA biosensors are usually utilized for identifying any single-nucleotide mismatches, DNA hybridization, and determining DNA interactions [45,46].

3.5.4   Tumor Identification

Many biosensors have been developed which can detect cancers and tumors in the human body. With the advent of *nanotechnology*, there is a gigantic impact on biosensors which have been exclusively developed for cancer detection, diagnosis, treatment, and regulation of the disease. *Nano* itself means small, but, fortuitously, the technology has been able to provide a humongous boost to efficient and robust biosensor systems for cancers [47]. Unfortunately, cancer is detected only after it becomes metastatic, referring to the spread in the other regions of the human body which, in turn, becomes tedious to diagnose and treat. Nanotechnology-based biosensors lucidly eliminate this loophole and improve the detection of metastatic cancers and also the survival rate of the patient. The employment of nanomaterials, such as liposomes, dendrimers, buckyballs, and nanotubes, as imaging agents aids for more sensitive and accurate quantification of cancerous tissues [48]. Such small-sized yet efficient biosensors are affordable, easy to use, patient-friendly, helps to access and identify the cancer markers, and remarkable detection in the first place. These biosensors have been widely appreciated by both the consultants and patients [49,50].

3.5.5   Medicine dosage management

Personalized medicine is the new trend today which states the four postulates of living a healthy way of life—*right drug* of the *right dosage* to the *right person* at the *right time* [51]. People are very conscious now of the type of medications that they intake. In order to manage the drug dosage and its intake, investigators have settled for a medicine adherence biosensor which looks no less than a capsule. RFID (radio-frequency identification)-tagged gelatinous capsule is one of those biosensors which is preferred by many toxicologists and consultants. It is the best monitor for drug adherence, affinity to targets, and any toxicology or side effects in the body. It gets activated only when it gets dissolved in the stomach as the RFID tag then sends a signal to a relatable device which propagates time-bound information to a cloud-based server which acts as an exact quantification of the medication adherence. It has the capacity to maintain connectivity and interactivity with the user. The only hesitation of this biosensor with patients is the fact that the biosensor has metallic compounds in its composition, which can be harmful to the intestines. However, more than limitations this biosensor has advantages. With a few modifications, this biosensor can be easily employed in methodical management of medicines. Moreover, consultants can also use the biosensor for monitoring patient behavioral pattern, if any side effects or poisoning with the medicine, rate of efficacy of the drug, etc. [52].

3.5.6   ECG regulation

Generally, wearable and implantable devices function well when used externally [53]. ECG devices were also based on the same principle previously where they were functional only when the power supply was fed externally [54]. These were completely perfect for patients who could move easily and had no serious problems. But, for patients whose movement was restricted, such as people with arrhythmia, it becomes mandatory to check for pivotal signs all round the clock and cannot be done using the traditional ECG machines. In order to resolve the issue, Lee and Seo [55] proposed the wireless power transfer technology in order to charge the ECG device wirelessly. This utilization of wireless charging technology entirely removes the use of external wired power charging of ECG devices, giving power transfer efficiency of approximately 30%. Another bandage-styled biosensor known as WiSP is affordable, lightweight and capable of energy harvesting, and can be connected to a smartphone using near-field communication (NFC) for use in both hospital and home-based settings [56]. Both biosensors have shown best results in monitoring cardiac and heart rate of patients.

3.5.7   Menstruation and fertility maintenance

Female fertility monitoring and management is another domain which is rapidly gaining attention in our modern times. There are many web-based sensors and biosensors which are being used religiously by females all over the globe. The device has been designed in such a manner which can easily estimate the fertile and infertile periods in female menstrual cycle which is based on the alterations in the amount of human chorionic gonadotropin (hCG) hormone luteinizing hormone (LH) in the urine. Based on this principle, currently there are four types of biosensor systems for checking for pregnancy and normal menstruation in females, namely Luteinizing hormone monitors (Clearblue Easy, Persona), Thermal monitors (DuoFertility, Lady-Comp, Baby-Comp, Pearly), Electrolyte monitors (OvaCue and OvWatch), and Cyclothermal monitor (Cyclotest), respectively [57].

3.5.8   Ocular and skin disorders

Early and fast diagnosis and evaluation of diseases is gaining momentum in the general public as it is important for maintaining a healthy lifestyle and also helps the patient to check on their diseases. Till now, we have already encountered many biosensors which were either dependent on urine, saliva, blood, skin, fluid discharge, etc. Tear fluid is not only helpful in ocular diseases, but also for checking for diabetes and cancers. It is only because of its noninvasiveness and least complexity. Ocular-based biosensors function by observing the alterations in the analytes in tear discharge, which in turn helps to determine the progress of the disease. Tear discharge is highly enriched with proteins similar to blood [58]. Because of this similarity between tear discharge and blood, biosensors are developed which specifically target analytes only within the tear. Contact lenses are not only meant to embellish the eyes with different beautiful colors but also give an open platform for addressing to ocular issues and for their diagnosis. They have been successfully affiliated with the management and maintenance of diseases such as diabetes and glaucoma [59]. The popularity of lacrimal contact lens biosensors is so high that big companies, Google and Novartis, started a new line of contact lens biosensors for diabetes regulation [60]. Apart from eyes, there are many biosensor systems which have been exclusively developed for skin-related problems.

TV commercials often advertise that one must leave the house by first applying a thick layer of sunblock or sunscreen, or there are also advertisements for dry skin which usually is experienced in the winters. No matter what the season, one must always take good care of their skin. There is an adage which goes like *Beauty is Skin deep!*, where the literal meaning refers to the fact that external beauty is not worthy enough when compared to one's inner beauty and serenity. A good healthy skin shows ones good and healthy lifestyle. UV radiations from the sun can harshly damage our skin leading to some very serious problems such as carcinomas, melanomas, skin patch and patterns, hyperpigmentation, etc. In order to prevent such skin problems, many researchers have amalgamated immunological analyses with artificial neural networking (ANN), deep convulsional neural networking (DCNN), and various other machine learning approaches. Schmidt and Zillikens [61] postulated a skin diseases detection system based on the study of autoantibodies and mucous membranes by the employing immunofluorescence (IF) microscopy. Esteva et al. [62] gave a deep learning approach in order to classify skin lesions, while, on the other hand, Codella et al. [63] proposed a method for skin cancer detection which was also based on deep learning strategy. A latest smart biosensor system has been developed by Połap et al. [64] which includes a homely-based smart camera which captures images of the skin and hunts for any severely damaged skin alterations using a key point search strategy. Once the image gets clustered as per their locations, the chosen clusters are forwarded to the neural network model which then analyses them for skin features. This biosensor system is simple and dynamic in nature. It can give the slightest skin changes as outputs on the connected device the user chooses. The outputs of this model have accuracy rates of around 80%–82.4%.

4   Advantages and limitations of *smart biosensors* in PoC well-being

In the past few decades, the biosensor technology has grown from simple and cheap components to the integration of multiple sensors into a single unit, making it smaller, cheaper, and tailored for mass production. The immense application of biosensor in medical diagnostic has driven scientists and engineers in the evolution of biosensor technologies. Some of the important advantages of biosensors in sensing a variety of molecules in medical diagnostic are: (1) simple to use and operate, (2) high sensitivity, (3) capable of doing multiple analyses, and (4) allow integration with different functions by same chip [65]. The high sensitivity and selectivity allow diagnosing the targeted disease at early stages. For the real-time monitoring, analysis, prediction of health condition and medical decision-making, artificial intelligence (AI), machine learning, big data analytics, and the internet of things (IoT) have also been integrated with biosensors. These analytics tools also allow real-time monitoring of biomarkers (e.g., tumors) which help doctors understand the epidemic and progression under therapy [66]. Recently, implantable biosensors have been introduced as the next-generation biosensors for personalized healthcare.

The advancement in nanotechnology and lab-on-chip-analysis systems adds further feathers to biosensor technology, providing microfabrication, homogenous sensing format, etc. However, its cost needs to be minimized to make it more affordable to all groups of people.

After the invention of Clark's electrode during the 1950s, a significant milestone has been covered in the field of biosensor technologies during last 7 decades. However, its practical applications in the field of medical science are still in its infancy. The biosensing devices need further improvements in order to make it a more precise diagnostic tool. The improvements can be in terms of simplicity, affordability, sensitivity, specificity, accuracy, multiplex analysis of multiple biomarkers, reduction in reagent consumption, reusability of biological sensing elements, making the device configurable for continuous monitoring, and so on. Since the ultimate goal of biosensor devices is to provide bedside PoC testing, therefore, further development in the technology such as multiplexing ability, fabrication, and miniaturization of biosensors, and so on are necessary in order to provide lab-on-chip-analysis to the community by replacing time-consuming laboratory analysis in medical diagnostics.

Recently, biosensors are being developed for cancer-related clinical testing with improved ease of use and error-free analysis of tumors. Also, biosensors have opened up new opportunities for exploring genetic signatures of the tumor profile in cancer testing. The successful application of biosensors in PoC molecular testing requires (1) ultrasensitive transducers, (2) changeable biorecognition elements, (3) integration, (4) miniaturization, (5) automation of sample preparation and amplification steps, (6) need of low sample and reagent volume, and (7) lower cost. Development in these features is the major hurdle for the rapid growth of these technologies. Table 4.1 summarizes the advantages and disadvantages of biosensors [65–67].

Table 4.1 Advantages and disadvantages of biosensors.

Advantages	Disadvantages
Simple to operate	Expensive
High sensitivity	Requires maintenance
Multi processing capability	Large size difficulties
Highly parallel in nature	—
Many functions on one chip	—
Real-time monitoring	—

5 Past ventures of biosensors

We have almost come to the end of this chapter and it seems every aspect of biosensors and their *smartness* has been discussed. This section highlights the history of biosensors, their current status, and how can their performance be improvised in the near future. *Leland Clark* is known as the "Father of Biosensors" as he developed and introduced the very first working biosensor for oxygen detection in 1956, which was then named as The Clark Electrode [8]. However, biological sensors were first worked prior to this in the year 1906 by M. Cremer where he illustrated the concentration of an acid in a liquid is directly proportional to the electric potential which develops between parts of the fluid present on opposite sides of a glass membrane [68]. Although the experiment was based on the preliminary mechanism of a biosensor, it came to limelight in 1909 with the help of Lauritz Sørenson's hydrogen ion concentration (pH) theory. Hughes in 1922 discovered pH measurement-based electrode which made the idea of a biological sensor even more realistic and vivid [69]. As years passed, in 1962, based on the same concept, Guilbault and Montalvo in 1969 proposed the first potentiometric biosensor to determine urea [70]. After this, commercialization and industrialization got better and expanded. A company named Yellow Spring Instruments developed the first marketable biosensor in 1975.

6 Present availability for biosensors

Since then, the market of biosensors has never looked back and, currently, it is an amalgamation of many scientific fields such as physics, chemistry, biology, microbiology, nanotechnology, medicine, electronics, and computational sciences [8]. Wide applications are in different fields that surround smart biosensors today, some of which have already been described in this chapter, which mainly correspond to drug discovery and design, food safety and quality monitoring, security, and PoC management, environmental monitoring, forensic sciences, agricultural sciences, disease prediction and biomarker detection, etc. Biological sensors have seen a tremendous transformation in their size, functioning, design, and programming. From simple acid concentration changes to the use of nanoparticles, the smartness of biosensors has increased from past ages. From glucometers using electrochemical identification of oxygen to hydrogen peroxide being used to immobilize glucose oxidase extrude has unleashed improvement in biosensors for making them more efficient and up-to-date with current time. Additionally, wet-lab experimentations such as fluorescence have also been used to prepare a biosensor [67]. Nowadays, a nanomaterial is labeled with a fluorescent tag for much higher specificity of biosensors. More innovative biosensors have been developed by using short stretches of aptamers, peptides, nucleotides, molecule imprinted polymers, etc. For bringing novelty to biosensors, it is mandatory to bring multiple fields to one platform so as to provide myriad applicability to biosensors. Moreover, this also helps to increase the flexibility of use of such electronic biosensors [71].

7 Future prospects for biosensors

Although biosensors have trotted a long route from 1909 to 2019, there are still some modifications which can be incorporated. Medical and healthcare applications still lag behind with only a few applications such as glucometers, ECG management, cancer prediction, biomarker identification, female hygiene, and blood pressure monitoring. There is an urgent need for the development of more PoC diagnosis and prognosis-based biosensors which can help consultants to diagnose diseases as early as possible. As far as our understanding goes, we suggest a biosensor should be highly *specific, sensitive, selective, multiprocessing, highly parallel in nature*, and *miniature-sized* [72]. Only then, we can hope that successful biosensors in the future will hold the potential to entirely change the current frame of medical code of conduct by detecting even the smallest targets within a short frame of time rapidly. Developments of such biosensors are highly needed and would be even more appreciated by everyone.

8 Discussion

Cammann [73] was the first one who phrased the term *biosensor*, whereas Leland Clark is known as the *Father of Biosensors* as he developed and introduced the very first working biosensor for oxygen detection in 1956. These are simply a combination of two major components—*biological* and *sensors*. Biosensors are flexible devices with appropriate selectivity and sensitivity, holding the potential to convert biological information into important physical/chemical or electrical signals, namely based on the receptor–transducer property. These devices are usually *wearable* making it easy for people to maintain their healthcare routine. The most common and commercial wearable devices these days is the "FitBit" activity tracker band which monitors the number of steps trotted by the individual, heart rate, sleep quality, steps climbed, and other personal measures engaged with fitness; some glucometers such as Accucheck and Dr. Morepen blood glucose monitors are another sort of biosensors which are used to monitor the glucose level in diabetes patients. Today, with the rise of healthcare awareness, many industrial and IT companies have launched biosensors for almost every healthcare-related problem. With the advent of IT, a lot of development with respect to PoC diagnosis and prognosis-based biosensors can help consultants to diagnose diseases as early as possible. A biosensor should be highly *specific, sensitive, selective, multiprocessing*, and *highly parallel in nature and small*-sized which can change the current frame of medical code of conduct by detecting even the smallest targets within a short frame of time rapidly.

9 Conclusion

A biosensor system is mainly composed of three components: (1) a recognition region (bioreceptor), (2) a transducer element (biotransducer), and (3) an electronic circuitry which further has an inbuilt system of an amplifier, a processor, and a visualizing display. The architecture of a biosensor is simple, and its basic desideratum is to rapidly test and detect where the analyte was procured. A bioreceptor is a biomolecule which can successfully identify the biological analyte. A bioreceptor can either be an enzyme, tissues or cells, any of the nucleic acids, RNA or DNA, antibodies, etc. Once biorecognition event takes place, the physical/chemical signal from the bioreceptor gets passed onto the transducer which converts it into a measurable signal. This conversion of one form of energy signal to another by the biotransducer is termed as *signalization*. The interaction between the analyte and bioreceptor (biological signal) can either be in the form of light, charge, mass change, or heat. This electrical signal is then amplified and thereby visualized on the display. A biosensor's functioning is very much dependent on the type of biotransducer it has. There are mainly three kinds of biotransducers, namely *electrochemical*, *ion channel*, and *reagentless fluorescent biotransducers*. Based on sensor devices and biological material used, biosensors are classified as electrochemical, optical, electrical, piezoelectric, and thermometric.

The *smartness* of biosensors has wide implications which are presently being cherished not only by industrialists, academicians, and public but also by environmentalists, physicians/medical consultants, etc. *High specificity and sensitivity, reusability, affordability, easy usage*, and dependence of basic parameters make a biosensor *smart*. Biosensors have been applied in myriad domain, such as environmental studies, biotechnology and food industry, healthcare, marine domain, etc.

Since the past times when simple biosensors for simple uses were developed, with the advent of information technology, the market of biosensors has grown and expanded in many scientific fields such as physics, chemistry, biology, microbiology, nanotechnology, medicine, electronics, and computational sciences. Applications of biosensors have also been observed in drug discovery and design, food safety and quality monitoring, security, and PoC management, environmental monitoring, forensic sciences, agricultural sciences, disease prediction, and biomarker detection. Biological sensors have seen a tremendous transformation in their size, functioning, design, and programming. From simple acid concentration changes to the use of nanoparticles, the smartness of biosensors has increased from past ages. From glucometers using electrochemical identification of oxygen to hydrogen peroxide being used to immobilize glucose oxidase extrude has unleashed improvement in biosensors for making them more efficient and up-to-date with current time.

Biosensor technology has grown from simple components to the integration of multiple sensors into a single unit, and its immense application in medical diagnostic has driven scientists and engineers in evolution of biosensor technologies. Some of the important advantages of biosensors in sensing a variety of molecules in medical diagnostic are *simple to use and operate, high sensitivity, capable of doing multiple analyses*, and *allow integration with different functions by the same chip*. Artificial intelligence, machine learning, big data analytics, and the IoT have also been integrated with biosensors for real-time monitoring, diagnosis, prediction, and decision-making [74]. Also, implantable biosensors have been developed as the next-generation biosensors for personalized healthcare. However, the biosensing devices need further improvements in order to make it a more precise diagnostic tool. The improvements can be in terms of simplicity, affordability, sensitivity, specificity, accuracy, multiplex analysis of multiple biomarkers, reduction in reagent consumption, reusability of biological sensing elements, making the device configurable for continuous monitoring, and so on.

Although biosensors have experienced many changes from 1909 to 2019, there are still some modifications which can be incorporated for their better performances. Medical and healthcare applications still lag behind with only a few applications such as glucometers, ECG management, cancer prediction, biomarker identification, female hygiene, and blood pressure monitoring. There is a need for the development of more PoC diagnosis and prognosis-based biosensors which can help consultants to diagnose diseases as early as possible. A biosensor should be highly *specific*, *sensitive*, *selective*, *multiprocessing*, and *highly parallel in nature and miniature-sized*, which will hold the potential to entirely change the current frame of medical code of conduct by detecting even the smallest targets within a short frame of time rapidly. Efficient and smart biosensors development is ongoing and will be appreciated by everyone in the near future.

Acknowledgments

Sahar Qazi is supported by INSPIRE Fellowship of Department of Sciences & Technology, Government of India.

References

[1] S. Upadhyay, R. Sinha, Smart Biosensors in Healthcare. Health. Available from: http://www.scind.org/1319/Health/smart-biosensors-in-health-care.html, 2018.

[2] K. Raza, S. Qazi, Nanopore sequencing technology and internet of living things: a big hope for u-healthcare. Sensors for Health Monitoring, vol. 5, Elsevier, Academic Press, London, UK, 2019.

[3] Fitbit. Available from: https://www.fitbit.com/in/home, 2019.

[4] Dr. Morepen Glucose Monitor. Available from: https://www.drmorepen.com/products/dr-morepen-glucoone-bg-03, 2019.

[5] R. Burn, Biosensors, wearables and digital biotech, Chemistry World, The Royal Society of Chemistry, London, 2018.

[6] A. Hierlemann, O. Brand, et al. Microfabrication techniques for chemical/biosensors, Proc. IEEE 91 (6) (2003) 839–863.

[7] A. Hierlemann, H. Baltes, CMOS-based chemical microsensors, Analyst 128 (1) (2003) 15–28.

[8] N. Bhalla, P. Jolly, et al. Introduction to biosensors, Essays Biochem. 60 (1) (2016) 1–8.

[9] D.R. Thévenot, K. Toth, et al. Electrochemical biosensors: recommended definitions and classification, Biosens. Bioelectron. 16 (1–2) (2001) 121–131.

[10] A. Chaubey, B.D. Malhotra, Mediated biosensors, Biosens. Bioelectron. 17 (6–7) (2002) 441–456.

[11] D. Grieshaber, R. MacKenzie, et al. Electrochemical biosensors—sensor principles and architectures, Sensors (Basel) 8 (3) (2008) 1400–1458.

[12] P.B. Luppa, L.J. Sokoll, D.W. Chan, Immunosensors—principles and applications to clinical chemistry, Clin. Chim. Acta 314 (1–2) (2001) 1–26.

[13] A.P. Turner, Biosensors: switching channels make sense, Nature 387 (6633) (1997) 555, 557.

[14] B.A. Cornell, V.L. Braach-Maksvytis, et al. A biosensor that uses ion-channel switches, Nature 387 (6633) (1997) 580–583.

[15] C. Solscheid, M.R. Webb, Development of reagentless fluorescent biosensors for use in rapid kinetic assays, Biophys. J. 104 (2) (2013) 1:529A.

[16] R. Koncki, Recent developments in potentiometric biosensors for biomedical analysis, Anal. Chim. Acta. 599 (1) (2007) 7–15.

[17] J.G. Guan, Y.Q. Miao, Q.J. Zhang, Impedimetric biosensors, J. Biosci. Bioeng. 97 (4) (2004) 219–226.

[18] Y. Wang, Z. Ye, Y. Ying, New trends in impedimetric biosensors for the detection of food-borne pathogenic bacteria, Sensors 12 (3) (2012) 3449–3471.

[19] E.B. Bahadır, M.K. Sezgintürk, A review on impedimetric biosensors, Artif. Cells Nanomed. Biotechnol. 44 (1) (2016) 248–262.

[20] R. Monošík, M. Stred'anský, E. Šturdík, Biosensors-classification, characterization and new trends, Acta Chim. Slovaca 5 (1) (2012) 109–120.

[21] E.V. Korotkaya, Biosensors: design, classification, and applications in the food industry, Foods Raw Mater. 2 (2) (2014) 161–171.

[22] H.M. Hiep, T. Endo, K. Kerman, M. Chikae, D.K. Kim, S. Yamamura, … E. Tamiya, A localized surface plasmon resonance based immunosensor for the detection of casein in milk, Sci. Technol. Adv. Mater. 8 (4) (2007) 331–338.

[23] X. Luo, J.J. Davis, Electrical biosensors and the label free detection of protein disease biomarkers, Chem. Soc. Rev. 42 (13) (2013) 5944–5962.

[24] P. Mehrotra, Biosensors and their applications—a review, J. Oral. Biol. Craniofac. Res. 6 (2) (2016) 153–159.

[25] N. Verma, M. Singh, Biosensors for heavy metals, Biometals 18 (2) (2005) 121–129.

[26] V. Scognamiglio, F. Arduini, et al. Bio sensing technology for sustainable food safety, Trends Anal. Chem. 62 (2014) 1–10.

[27] M. Ghasemi-Varnamkhasti, M.L. Rodriguez-Mendez, S.S. Mohtasebi, Monitoring the aging of beers using a bioelectronic tongue, Food Control 25 (2012) 216–224.

[28] C. Ercole, M. Del Gallo, et al. *Escherichia coli* detection in vegetable food by a potentiometric biosensor, Sens. Actuat. B Chem. 91 (2003) 163–168.

[29] C. Yan, F. Dong, et al. Recent progress of commercially available biosensors in china and their applications in fermentation processes, J. Northeast Agric. Univ. 21 (2014) 73–85.

[30] R. Mishra, R. Dominguez, et al. A novel automated flow-based biosensor for the determination of organophosphate pesticides in milk, Biosens. Bioelectron. 32 (2012) 56–61.

[31] L. Su, W. Jia, et al. Microbial biosensors: a review, Biosens. Bioelectron. 26 (5) (2011) 1788–1799.

[32] S. Daunert, G. Barrett, et al. Genetically engineered whole-cell sensing systems: coupling biological recognition with reporter genes, Chem. Rev. 100 (7) (2000) 2705–2738.

[33] K.M. Narayan, J.P. Boyle, et al. Impact of recent increase in incidence on future diabetes burden: U.S., 2005-2050, Diabetes Care 29 (2006) 2114–2116.

[34] E.H. Yoo, S.Y. Lee, Glucose biosensors: an overview of use in clinical practice, Sensors (Basel) 10 (5) (2010) 4558–4576.

[35] G.H. Murata, J.H. Shah, et al. Intensified blood glucose monitoring improves glycemic control in stable, insulin-treated veterans with type 2 diabetes: the diabetes outcomes in veterans study (DOVES), Diabetes Care 26 (2003) 1759–1763.

[36] L.G. Jovanovic, Using meal-based self-monitoring of blood glucose as a tool to improve outcomes in pregnancy complicated by diabetes, Endocr. Pract. 14 (2008) 239–247.

[37] N. Poolsup, N. Suksomboon, S. Rattanasookchit, Meta-analysis of the benefits of self-monitoring of blood glucose on glycemic control in type 2 diabetes patients: an update, Diabetes Technol. Ther. 11 (2009) 775–784.

[38] S. Skeie, G.B. Kristensen, et al. Self-monitoring of blood glucose in type 1 diabetes patients with insufficient metabolic control: focused self-monitoring of blood glucose intervention can lower glycated hemoglobin A1C, J. Diabetes Sci. Technol. 3 (2009) 83–88.

[39] E.I. Boutati, S.A. Raptis, Self-monitoring of blood glucose as part of the integral care of type 2 diabetes, Diabetes Care 32 (Suppl. 2) (2009) S205–S210.

[40] S.L. Tunis, M.E. Minshall, Self-monitoring of blood glucose (SMBG) for type 2 diabetes patients treated with oral anti-diabetes drugs and with a recent history of monitoring: cost-effectiveness in the US, Curr. Med. Res. Opin. 26 (2010) 151–162.

[41] E. Ozkumur, et al. Label-free and dynamic detection of biomolecular interactions for high-throughput microarray applications, Proc. Natl. Acad. Sci. USA 105 (23) (2008) 7988–7992.

[42] O. Avci, N.L. Ünlü, et al. Interferometric reflectance imaging sensor (IRIS)—a platform technology for multiplexed diagnostics and digital detection, Sensors (Basel) 15 (7) (2015) 17649–17665.

[43] R. Rettner, DNA: Definition, Structure & Discovery, Live Science. Available from: https://www.livescience.com/37247-dna.html, 2017.

[44] J.M. Butler, Genetics and genomics of core short tandem repeat loci used in human identity testing, J. Forensic Sci. 51 (2006) 253–265.

[45] M.E. Minaei, M. Saadati, et al. DNA electrochemical nanobiosensors for the detection of biological agents, J. Appl. Biotechnol. Rep. 2 (1) (2015) 175–185.

[46] S. Reisberga, B. Piroa, et al. Selectivity and sensitivity of a reagentless electrochemical DNA sensor studied by square wave voltammetry and fluorescence, Bioelectrochemistry 69 (2006) 172–179.

[47] P. Grodzinski, M. Silver, L.K. Molnar, Nanotechnology for cancer diagnostics: promises and challenges, Expert Rev. Mol. Diagn. 6 (3) (2006) 307–318.

[48] B. Bohunicky, S.A. Mousa, Biosensors: the new wave in cancer diagnosis, Nanotechnol. Sci. Appl. 2011 (4) (2011) 1–10.

[49] A. Rasooly, J. Jacobson, Development of biosensors for cancer clinical testing, Biosens. Bioelectron. 21 (10) (2006) 1851–1858.

[50] C. Costa, M. Abal, et al. Biosensors for the detection of circulating tumour cells, Sensors (Basel) 14 (3) (2014) 4856–4875.

[51] S. Qazi, Personalized medicines in psychiatry: promises and challenges, J. Appl. Comput. 2 (2) (2017) 50–55.

[52] P.R. Chai, J. Castillo-Mancilla, Utilizing an ingestible biosensor to assess real-time medication adherence, J. Med. Toxicol. 11 (4) (2015) 439–444.

[53] P. Gutruf, J.A. Rogers, Implantable, wireless device platforms for neuroscience research, Curr. Opin. Neurobiol. 50 (2018) 42–49.

[54] J.H. Lee, Human implantable arrhythmia monitoring sensor with wireless power and data transmission technique, Austin J. Biosens. Bioelectron. 1 (2015) 1008.

[55] J.H. Lee, D.W. Seo, Development of ECG monitoring system and implantable device with wireless charging, Micromachines (Basel) 10 (1) (2019) 38.

[56] S.P. Lee, G. Ha, et al. Highly flexible, wearable, and disposable cardiac biosensors for remote and ambulatory monitoring, Npj Digit. Med. 1 (2018) 2.

[57] AzoSensors, Fertility Monitoring Systems, AzoSensors. Available from: https://www.azosensors.com/article.aspx?ArticleID=38, 2012.

[58] M. Kohji, A. Takahiro, Cavitas sensors: contact lens type sensors & mouthguard sensors, Electroanalysis 28 (2016) 1170–1187.

[59] N.M. Farandos, A.K. Yetisen, et al. Contact lens sensors in ocular diagnostics, Adv. Healthc. Mater. 4 (6) (2015) 792–810.

[60] M. Senior, Novartis signs up for Google smart lens, Nat. Biotechnol. 32 (9) (2014) 856.

[61] E. Schmidt, D. Zillikens, Modern diagnosis of autoimmune blistering skin diseases, Autoimmun. Rev. 10 (2) (2010) 84–89.

[62] A. Esteva, B. Kuprel, et al. Dermatologist-level classification of skin cancer with deep neural networks, Nature 542 (7639) (2017) 115–118.

[63] N.C. Codella, Q.B. Nguyen, et al. Deep learning ensembles for melanoma recognition in dermoscopy images, IBM J. Res. Dev. 61 (2017) 5:1–:15.

[64] D. Połap, A. Winnicka, et al. An intelligent system for monitoring skin diseases, Sensors (Basel) 18 (8) (2018) 2552.

[65] S. Patel, R. Nanda, S. Sahoo, E. Mohapatra, Biosensors in health care: the milestones achieved in their development towards lab-on-chip-analysis, Biochem. Res. Int. 2016 (2016) 1–12.

[66] A. Kaushik, M.A. Mujawar, Point of care sensing devices: better care for everyone, Sensors (Basel) 18 (12) (2018) E4303.

[67] AzoSensors, Biosensor Technology: Advantages and Applications, AzoSensors. Available from: https://www.azosensors.com/article.aspx?ArticleID=402, 2013.

[68] M. Cremer, Über die Ursache der elektromotorischen Eigenschaften der Gewebe, zugleich ein Beitrag zur Lehre von den polyphasischen Elektrolytketten, Z. Biol. 47 (1906) 562–608.

[69] W.S. Hughes, The potential difference between glass and electrolytes in contact with the glass, J. Am. Chem. Soc 44 (1922) 2860–2867.

[70] G.G. Guilbault, J.G. Montalvo Jr., A urea-specific enzyme electrode, J. Am. Chem. Soc. 91 (8) (1969) 2164–2165.

[71] S. Vigneshvar, C.C. Sudhakumari, et al. Recent advances in biosensor technology for potential applications—an overview, Front. Bioeng. Biotechnol. 4 (2016) 11.

[72] S. Song, H. Xu, C. Fan, Potential diagnostic applications of biosensors: current and future directions, Int. J. Nanomedicine 1 (4) (2006) 433–440.

[73] K. Cammann, Biosensors based on ion-selective electrodes, Fresen. Z. Anal. Chem. 287 (1977) 1–9.

[74] S. Qazi, K. Tanveer, K. ElBahnasy, K. Raza, From telediagnosis to teletreatment: the role of computational biology and bioinformatics in tele-based healthcare, in: Telemedicine Technologies, Academic Press, USA, 2019, pp. 153–169.

Energy harvesting via human body activities

Sweta Kumari[a], Sitanshu Sekhar Sahu[a], Bharat Gupta[b], Sudhansu Kumar Mishra[a]

[a]*Birla Institute of Technology, Mesra, Ranchi, India;*
[b]*National Institute of Technology, Patna, India*

1 Introduction

Energy harvesting is the enabling technology which provides electricity for micro/ macro and portable equipment from different available energy sources such as solar energy, kinetic energy, thermal energy, wind energy, etc. Replacement of batteries from any electronic devices becomes very tedious when it has been deployed in remote location or in vivo. Energy harvesting is one of the attractive solutions for these types of problem and provides green and semipermanent supply of energy to the low power devices. Energy efficiency gradually increases and spreads their functionality in various fields by reducing their sizes. Therefore, harvesting energy from environment and human body activity makes breakthrough over decades. Majority of these applications are based upon the low frequencies and methods to be obtained for improved power outputs are different. To optimize these low frequencies, energy harvesting technology in terms of power conditioning circuits becomes a critical task. In the last decade, there is tremendous development in electronics, nanoparticles, and wireless sensor networks (WSNs) that enable a new generation of body-worn devices for healthcare [1–4]. Conventional wireless sensors are battery driven, so the lifetime of the node is short. Wireless sensors can be deployed in so many areas, such as bottomless sea [5], data distribution-based applications in adversity effect [6,7] controlling the adverse effect due to volcano [8], avalanche rescue [9], and underground monitoring [10] are few scenarios where replacement of battery is more tedious.

Therefore, the demand of a healthcare monitoring network with long life without replacement is present. Energy harvesting is one of the solutions for any wireless network. The conversion of an ambient energy from one form to other is known as "Energy Harvesting." There are many ways of gathering energies from available surrounding sources, that is, wind energy, solar energy, vibration energy, wastage, and chemical and biological energy [11]. The harvested energy is environment friendly, safe, economical and pollution free, and it also offers a feasible solution to the energy optimization problem. The rapid industrial developments make it imperative to figure out options of alternative sources of energy so that the dependence on conventional sources of energy can be reduced.

1.1 Wireless body area network and its applications

Wireless body area networks (WBANs) are multidisciplinary and intersectional technology for mobile healthcare industry and applicable for many applications, which has been shown in Fig. 5.1. WBAN is made of several small sensor nodes and a gateway which is connected to the sensor node within the range of telecommunication networks, such as mobile phones, WLAN, hospitals/medical care centers, or Wi-Fi [12]. It will allow the client to store the required data on his/her PDA (personal digital assistant), iPod, or laptop, and then transfer that medical information to a suitable medical server, that is, hospital or clinic. It will provide the suitability for remote patients' vital signs monitoring and diagnosis, thereby improving the quality of healthcare information monitoring [13,14] system. Hence, longevity of sensors in terms of lifetime is to be high for both use and maintenance and therefore reduces the power sources of conventional methods [15]. It basically provides the automatic medical care services to living body in every aspects of life and can be used in several scenarios, that is, in medical and nonmedical applications such as to help the disabled person, blind persons, ECG/EEG, glucose, hearing aid and blood sugar, etc., for healthcare monitoring purposes. It (IEEE 802.15.6 standard) [16] also plays an important role in the field of nonmedical applications, namely entertainment, sports, military, and surveillance.

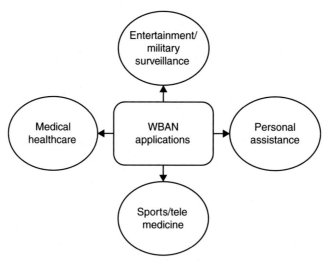

FIGURE 5.1 Wireless body area network (WBAN) and its applications.

Nowadays we are accessing 3G video calls through LED TVs or home audio system only by using this WSN technique [17] and also used for data file transfers, monitoring the lost things, military purposes, and social networking applications (that allow people to exchange their digital profile or business card only by handshaking). It becomes much more attractive, when the power sources come from environment or human bodies because of its appealing features.

1.2   Energy harvesting from human body

Fig. 5.2 depicts brief sketch of energy harvesting from human body. It takes the energy from different human body parts using the appropriate energy sources and then it can power the wireless sensor nodes and hence it is applicable for healthcare systems. Energy would be harvested from body motion, body heat, blood pressure, breathing, hand movements, pressing keyboards, perspiration, strolling and pedaling, etc. [18].

Wireless sensor mote consists of four blocks like sensor block, microcontroller (μC) unit, radio block, and power unit as shown in Fig. 5.2. To energize sensor mote in WSN, it is very important to adopt suitable methods so that all the components get sufficient power.

It requires sensors having distinguished features like portability, small size, lightweight, low power consumption, and autonomous sensor nodes that can monitor and control the health. The energy or power can be further utilized for powering the WBSNs motes for medical healthcare.

The major WBSN-based devices that are used in medical care are as follows [19]:

1. An electrocardiogram sensor (ECG) that monitors heart activity,
2. An electromyography sensor (EMG) that relies on muscle function activity,
3. An electroencephalogram (EEG) that monitors brain electrical activity,

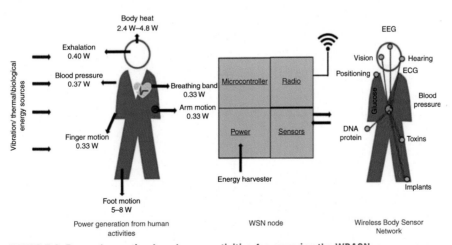

FIGURE 5.2 Energy harvesting from human activities for powering the WBASN.

4. A blood pressure sensor that monitors the extra force exerted on walls of blood vessels,
5. A breathing sensor used for monitoring the respiration,
6. Motion sensors that are used to distinguish the level of activity,
7. Temperature sensor that controls the variation of temperature between human body and surroundings, and
8. A pulse oximeter that measures the oxygen saturation levels in the human's blood.

Section 2 gives the brief analysis of energy harvesting module and principle behind the energy harvesting concept, Section 3 describes energy potential of a human body from several activities, Section 4 depicts the different energy harvesting techniques with their basic details, Section 5 suggests the several methods of extracting energy and their power consumptions, and Section 6 gives the brief details of biomedical implants and their power consumption.

2   Basic blocks of energy harvesting

A simple block diagram of energy harvesting system is depicted in Fig. 5.3. It consists of AC-DC rectifier circuit (convert AC into DC signal of energy sources), voltage regulator (gives fixed level of voltage and current), DC-DC converter (provides a boost-up output as input is very low), and temporary storage device (which actually stores the electricity) as subblocks.

Human body is busy with numerous activities which generate huge amount of mechanical energy that includes muscle stretching, arm swings, walking, running, heart beats, and blood flow [20]. Moreover, the behavior of these different movements is random in nature. So it is very much important to design a good and efficient energy harvesting circuit for better optimization in terms of battery lifetime, low power consumptions, and the desired voltage levels for wireless sensors.

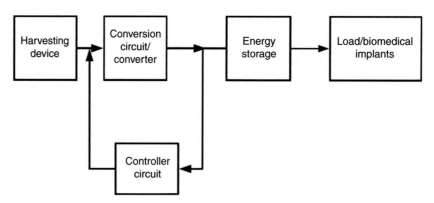

FIGURE 5.3 Basic blocks of energy harvesting system.

3   Basic energy harvesting principle

As energy cannot make and destroy, only convert from one form to another, this basic concept gave astounding effect to this era. Energy harvesting is made of several blocks having their specific tasks toward the input and output sources. There are three main ambient energy sources from human activities performed through body. They are mechanical energy (vibrations, deformations), thermal energy (temperature gradients or variations), and chemical energy (chemistry, biochemistry) [21].

3.1   Kinetic energy

Kinetic energy harvesters provide electrical energy from different human body motions [11]. This kinetic energy can be categorized into three parts, having different transduction methods: (1) electromagnetic, (2) electrostatic, and (3) piezoelectric.

3.2   Biological fuel cells (bio-FCs)

Biological fuel cell (bio-FCs) is transforming chemical energy into electrical via electrochemical reactions involving biochemical pathways [22]. The bio-FC can be divided into two parts, that is, bioelectrochemical systems (BES) and enzymatic fuel cells (EFC), which depend upon the catalyst used in the system for a living cell (often bacteria) or enzyme.

3.3   Thermal energy

Thermoelectricity is the phenomena of creating potential drop between two different metals or semiconductors and hence a seeback effect appears [23]. And for getting the better result or boosted output of circuits, it is very much important for opting appropriate input sensors.

4   Energy potential of a human body from several activities: motivation

Tremendous development in electronics era made our life technology dependant. In other words we can say that increasing demand of electronic gadgets such as mobile phones, global positioning system (GPS), laptops, etc., made us power hungry. And these power resources will fulfill only if we can recharge our battery in regular interval (i.e., high lifetime) or making battery less devices. For this approach we have to concentrate on designing of energy harvesting circuits with high efficiency and better voltage output as per the requirements.

Table 5.1 gives brief analysis of different energy scavenging from human body, such as using biofuels (extracts energy from biochemical reactions), mechanical movements (piezoelectric harvesters), body temperature variations (thermoelectric

Table 5.1 Energy harvesting from human body environment.

Human body acts as a power source	Device (EH system)	Functionality
Biofuel	Biochemical harvester	Extracts energy from biochemical reactions
Heat	Thermoelectric harvester	Generates power from temperature differences between two surfaces
Light	Photovoltaic harvester	Generates electricity from photons of light
Strain	Biomechanical harvester	Converts biomechanical strain/pressure into electricity
Vibrations	Piezoelectric harvester	Extracts energy from mechanical movements

harvesters), mechanical pressures (biomechanical harvesters), and light (photovoltaic harvesters).

There are several ways of scavenging energies from human body activities such as from thermoelectric methods, piezoelectricity, photovoltaic harvesters, and biomechanical harvesters [24], etc., which have been depicted in Table 5.2. It gives the different methods of energy harvesting with their electric equivalent circuits and their pros and cons.

Table 5.2 Human energy expenditure from different activities [25,26].

S. no.	Activity	Kcal/h	Watts
1.	Sleeping	70	81
2.	Lying quietly	80	93
3.	Sitting	100	116
4.	Standing at ease	110	128
5.	Conversation	110	128
6.	Eating a meal	110	128
7.	Strolling	140	163
8.	Driving a car	140	163
9.	Playing violin or piano	140	163
10.	Housekeeping	150	175
11.	Carpentry	230	268
12.	Hiking, 4 mph	350	407
13.	Swimming	500	582
14.	Mountain climbing	600	698
15.	Long-distance running	900	1048
16.	Sprinting	1400	1630
17.	Body heat [27]	0.118	130 μ

5   Human energy expenditure from several activities

Human bodies have enormous amount of energy sources which can be converted into electricity by the use of different harvesting methods. Table 5.2 shows the different activity of human and its power consumption. 116 W of power is generated from sitting posture, similarly 128 W of power is observed while eating meals [25, 28].

    Human does many things in daily life scenario, namely sleeping, breathing, walking, eating meals, and so many activities which are depicted in Table 5.3. All activities produce loads of energy and hence this will be considered as wastage or no use. Energy harvesting is the technology which can utilize these wastages by using appropriate methods and their applications.

6   Methods of extracting energy from body activities and their power consumptions: literature review

6.1   Kinetic energy harvesting methods

The power extracted from human gait can be measured in several ways. These power sources from human body activities are of several forms, such as heel strikes, shoulder movements, finger movements, center of mass motions, etc. For getting the appropriate power consumption of each activity, we are considering and performing an integrating analysis of different data of activities. For example, calculation of upper body part movements can be done in two ways: (1) through displacement and (2) concept of torque and angular movement. This is to be demonstrated by Eqs. (1) and (2).

$$W = \int_0^s F.dS \tag{1}$$

$$W = \int_0^\theta \tau.d\theta \tag{2}$$

where F and τ denote the force and torque, S and θ represent linear and angular displacements, respectively [38].

    Power from breathing:

$$W = p\Delta V \tag{3}$$

where W is maximum available power, p is pressure exerted from respiration. And ΔV is volume consumed by respiratory system.

    Power calculation has been depicted by respiration rate of 10 breaths/min and force exerted by 0.05 m distance is 100 N as follows:

$$(100\ N) \times (0.05\ m) \times (10\ breaths/1\ min) \times (1\ min/60\ s) = 0.83\ W$$

There are several ways of extracting power from intended or unintended sources, such as power from typing, from arm motion, from respiration, and from cycling and walking.

Table 5.3 Several energy harvesting methods.

Harvesting methods	Description	Equivalent circuit	Basic block	Merits	Demerits
1. Kinetic energy	Produces electrical energy from different human body motions	—	—	—	—
a. Electromagnetic devices [29]	Relative motion between a coil and a magnet generates electricity	—	[29]	Increased output currents, long life span, and robustness	Low output voltages, hard to develop MEMS devices, expensive, low efficiency, and small sizes
b. Electrostatic devices [29]	From the variable capacitor of two plates		[29]	High output voltage, adjustable coupling coefficient, miniature, and economic	Low capacitance, needs to control µm dimensions, and no direct mechanical-to-electrical conversion for electret-free converters
c. Piezoelectricity [30,31]	Conversion of mechanical into electrical energy and vice-versa by piezoelectric materials	[30,31]		Frequency response, self-start-up, dynamic ranges, and easy to use, high output voltages and capacitances No need to control any gap	Needs high impedance cable for getting active, coupling coefficient linked to material properties

Harvesting methods	Description	Equivalent circuit	Basic block	Merits	Demerits
2. Thermoelectricity [23,32]	Potential drop between two different metals or semiconductors and hence a seeback effect appears	[23,32]		High scalability and lower production cost	Low energy conversion efficiency rate, slow technology progression, requires relatively constant heat source
3. Enzymatic biofuel cell [33–36]	Break down the high energy liquids or body fluids such as blood, protein powder, sweat, and tears, and produce electricity thereby	[33–36]		Used in biomedical implants due to glucose oxidant as a source of energy	Short lifetime and low efficiency Low power density
4. Triboelectric Motion Sensor [37]	Convert mechanical response into electric fluctuation due to sliding or contacts of materials on each other	[37]		Used in energy harvesting technology with low cost with high output voltage in low random motion	Low efficiency

6.2 Biochemical energy harvesting methods

Enzymatic biofuel cell provides power to the biomedical implants as well as consumer electronics for better operation having simple and small miniaturizing device. It is specifically working in the blood vessels and after dissolving of glucose it makes fuel and oxidant [39,40].

$$\text{At anode: } C_6H_{12}O_6 \rightarrow C_6H_{10}O_6 + 2H^+ + 2e^- \tag{4}$$

$$\text{At cathode: } O_2 + 4H^+ + 4e^- \rightarrow 2H_2O \tag{5}$$

The electrochemical glucose biosensors are simple, flexible, and low cost of electrochemical transduction instrumentation properties that make these sensors devices the most commercial use for healthcare and many more applications.

6.3 Thermal energy harvesting methods

Human body food intake gives only 15%–30% [41] of mechanical efficiency by converting chemical into electrical energy that shows the maximum energy consumed by food releases into the environment and it makes very essential to utilize this energy into useful one.

Energy from thermoelectric generator converts heat energy (Q) into electrical, η is the efficiency term, and P denotes the thermoelectric power.

$$P = \eta * Q \tag{6}$$

$$\Delta T = \text{Th} - \text{Tc} \tag{7}$$

$$V = \alpha \Delta T \tag{8}$$

Eqs. (6–8) show the basic mathematical representation of thermoelectric generator.

It indicates that the maximum power is possible by using the electromagnetic generators in the range of milliwatts to several watts by using different piezoelectric transducers while using dielectric elastomers will be in the range of tens of μW to pW [42]. If we analyze the thermoelectric generator that depends upon the environmental temperature differences, it lies between hundreds of microwatts to a few milliwatts [42].

6.4 Power consumption and data rates of IMDs

Fig. 5.4 gives the basic information of different methods of biomedical implanted devices and their power ranges used for these [43].

Table 5.3 gives the several energy harvesting methods and their basic equivalent circuit analysis with advantages and disadvantages. It gives the overview of energy harvesting devices for several applications. Table 5.4 depicts the power consumption from different human body activities and their respective scavenging methods. Table 5.5 gives the brief analysis of data rate of few biomedical parameters which helps them adopting an appropriate energy harvesting technology.

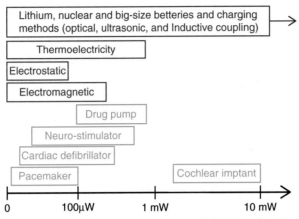

FIGURE 5.4 A rough scale of power ranges of methods and implanted biomedical devices [43].

Table 5.4 Power consumption from different human body sources and their respective harvesting technologies [44].

S. no.	Source	Harvesting method	Power
1.	Foot	Rotational electromagnetic harvester in heel	1.8 W
		Linear EM harvester in shoe	838 mW
		Piezoelectric in heel	8.4 mW
		Piezoelectric in sole	1.3 mW
		Electrostatic on ankle	61 pW
2.	Knee	Electromagnetic deceleration only	4.8 W
		Piezoelectric pinwheel	7 mW
		Dielectric elastomer	25 µW
3.	Torso	Spherical electromagnetic	1.44 mW
		Linear electromagnetic	2.46 mW
		Thermoelectric	4 mW
4.	Arm	Impact-based piezoelectric	40 µW
5.	Chest	Respiratory rotary electromagnetic	15 mW
6.	Wrist	Thermoelectric	250 µW
7.	Heart beat [27]	Thermoelectric	130 µW

Table 5.5 Biomedical parameters and its data rate [19].

Medical device	Data rate (Kbps)
ECG (12 leads)	144
EMG	320
EEG	43.2
Blood saturation	16

7 Analysis of power consumption from different sensor devices

Table 5.6 depicts the brief analysis of few medical devices based upon multiple harvesting techniques and their respective power consumptions. The cochlear battery sensor chip has been developed by researchers of MIT; it can harness the electrical energy without interfering with normal hearing and monitors biological activities in ears having imbalance impairments. UK researchers from Cranfield University suggested knee-joint piezoelectric harvesters which use plucking techniques for frequency up-conversion and used to power the body monitoring devices like heart rate monitors, pedometers, and accelerometers [45].

Riga Technical University, in Latvia, researchers have demonstrated a mechanical energy harvester while walking that has a planar structure and, using electrodynamics converter, during unintended human motions, it generates voltage pulses due to relative motion between motion and generator. Chinese–US research team proposed an elastomer-based piezoelectric fabric which generates electricity and illuminates 30 LEDs while walking. When the same fabric was applied onto a shirt and by artificial movement, it rapidly charged a lithium-ion battery [50]. The University of Waterloo (UW), in Canada, researchers are working on a wideband hybrid energy harvester which actually prolongs the battery life, reducing the several times of heart surgeries [51]. The novelty of this design is that it converts ambient vibrations into electricity using a combination of smart materials in order to operate at a wider range of frequencies. If the rate of motion of the vibration source decreases, so does the frequency, and the level of energy produced. A modern pacemaker operates on approximately 0.3 µW and these patches generate up to 0.18 µW/cm^2, when attached to the right ventricle of the heart [60]. Nadim Inaty designed a green wheelby that transforms kinetic energy produced by the human body into electricity. In a 30 min of jogging, it harvest the stored kinetic energy and produce enough power to charge the laptop for 2 h and lighting the bulb for 5 h [52]. Ref. [53] eliminates the dependency upon batteries for instrumented knee implant. A piezoelectric energy harvester has been embedded into knee implant and gets constant power. Ref. [56] gives the tremendous analysis of using helix pattern of piezoelectric harvesters, used in low frequency of 4 Hz with high output voltage of 140V P-P, 0.6 mW. Hybrid structure of piezoelectric and electromagnetic energy harvester gives tremendous result of 5.76 mW which can be easily used in biomedical implants [57]. Ref. [58] gives average power of 58.06 mW by using clicking mechanism piezoelectric energy harvester and Ref. [59] shows the power consumption of device is 0.333 mWh by using one shot pulse boost converter circuit using synchronized switched harvester on inductor (SSHI) piezoelectric energy harvester. Wireless ECG monitoring system using Zigbee technology has been presented in [60] which periodically monitors the patients in their home itself. Ref. [61] depicts about internet of things (IoT) which gives facilities such as sensing, processing, and communicating with its basic parameters. Ref. [61,62] concentrate on IoT combined with big data analysis. Medical cyber-physical systems (MCPS)

Table 5.6 Analysis of some medical devices and their power consumptions.

References	Type of devices	Methods	Power consumption	Year	Applications	Device images
[46]	Cochlear sensor battery	Endocochlear potential	1.12 nW from the EP for 5 h	2012	Power sources for hearing aid	
[47]	Fully implantable cochlear SoC battery	Piezoelectric sensors	More than 1.12 nW	2014	Monitor biological activities in ears, therapies	
[48]	Jacket with electrical generator	Based on electrodynamics converter which consists of the set of flat, spiral-shaped coils, and a block-shaped magnet	0.2 mW (power density of generator is about 1.8 mW/cm^3)	2011	Wearable electronics, energy per unit mass in batteries is limited	
[45]	Pizzicato knee-joint harvester	Piezoelectric EH. Using plucking technique	1.03 ± 0.08 mW	2012	Heart rate monitors, pedometers, and accelerometers	
[38]	AIRE mask	Electronic mask contains tiny wind turbines	—	2012	Run IPOD and mobile phones	

(Continued)

Table 5.6 Analysis of some medical devices and their power consumptions. (*Cont.*)

References	Type of devices	Methods	Power consumption	Year	Applications	Device images
[49]	Chin strap harvester circuit	PFC chin strap head-mounted device that harvests energy from jaw movements	Max. o/p power was 18 µW, optimum power o/p is 7 µW	2014	Power is too low but multilayer PFCs made it several applications like charging battery	
[50]	Glucose EFC	Carbon nanotube-based EFC	40 µW	2012	LED and digital thermometer	
[51]	Flexible piezoelectric patch	Patch contains 500 nM PZT	0.18 µW/cm^2	2014	Pacemaker	
[38]	2-channel wireless EEG	Thermoelectric generator based	0.8 mW power	2007	Detect imbalance between the two halves of the brain, brain trauma, and its monitoring	
[52]	Human Hamster Wheel for Energy	Kinetic energy, 10 mm polycarbonate tube to offer energy	Running on Green Wheel for 30 min gives 120 W	2011	To power CFL bulb for 5 h, charging mobile 12 times, run a laptop for 2 h and desktop for 1 h	—
[53]	Instrumented knee implant	Four piezoelectric generators embedded with power conditioning circuit	59.4 mW	2017	Knee implants, consumer electronics	

References	Type of devices	Methods	Power consumption	Year	Applications	Device images
[54]	Handy motion-driven piezoelectric harvester	Frequency up conversion with PEH	175 µW	2015	Wearable sensor applications	
[55]	Air bladder turbine EH	Bladders embedded into shoes to induce airflow from foot strike	90.6 mW	2016	Smart watches, activity trackers, wearable cameras	
[56]	A highly stretchable piezoelectric helical	Two counter-wound helical structures with z-core structure made of an elastic fabric string	0.6 mW	2015	Wearable sensor application, highly deformable body parts.	
[57]	Piezoelectric and electromagnetic energy harvester	It uses combination of internal and external cantilever frequency up conversion	5.76 mW	2017	Wearable sensor applications	
[58]	Piezoelectric energy harvester using clicking mechanism	Electricity produces by hitting and vibrating cantilever beams	$P_{AVG} = 58.06\ \text{mW}_{RMS}$, $340.03\ \text{mW}_{PEAK}$	2016	Consumer electronics and medical applications	
[59]	Piezoelectric sensor device (finger tapping)	SSHI-based energy harvester using one shot pulse boost converter	$P_{cons} = 0.333\ \text{mWh}$, $P_G = 9.164\ \text{mWh}$	2018	Low power energy harvester applications	

also take part in medical healthcare system for getting continuous high-quality healthcare [63]. It also works on crucial issues in healthcare system in terms of efficiency and security.

8 Conclusion

This chapter presents the different techniques by which energy would be harvested by piezoelectric, thermoelectric, and electromagnetic devices especially for WBAN/BSN applications. Human energy expenditure from different activities and how do we use that energy as electric sources and their power consumptions have been discussed. A brief comparison of all the available energy harvesting technology and their power consumption for extending the battery lifetime has been presented here. Piezoelectric harvester model is the best way to scavenge energy among the all possible harvesting technologies from human body activities. As this harvester gives comparatively high power delivery of tens of microwatt to few milliwatts, this meets the necessary operating power consumption for biomedical implants. There should be interaction between the mechanical harvester and the power electronics in such a way that they are capable of mitigating the losses due to capacitive attributes for designing of any energy harvester system by improving the conversion efficiency due to resistive impedance across power converter. The finger tapping on piezoelectric device gives very low power consumption of approximately 0.33 mWh by power generation of 9.164 mWh by using one shot pulse SSHI-based boost converter. Enzymatic biofuel cells have many advantages over conventional energy devices because of the specific activity from enzymes and the capability of miniaturization for implantable medical devices such as pacemakers, defibrillators, glucose level monitors, etc. Harvesting energy from human activities especially in the field of medical/healthcare would be a vital solution and it will be more beneficial if internal circuitry of energy harvesting could be designed properly in the terms of efficiency and low power dissipation. This can be achieved by using highly efficient boost converter circuit with suitable controller inside the circuit.

References

[1] C.P. Price, L.J. Kricka, Improving healthcare accessibility through point-of-care technologies, Clin. Chem. 53 (2007) 1665–1675.

[2] D.-H. Kim, N. Lu, R. Ma, Y.-S. Kim, R.-H. Kim, S. Wang, J. Wu, S.M. Won, H. Tao, A. Islam, Epidermal electronics, Science 333 (2011) 838–843.

[3] J.A. Cramer, Microelectronic systems for monitoring and enhancing patient compliance with medication regimens, Drugs 49 (1995) 321–327.

[4] S.K. Sia, L.J. Kricka, Microfluidics and point-of-care testing, Lab Chip. 8 (2008) 1982–1983.

[5] J. Heidemann, M. Stojanovic, M. Zorzi, Under water sensor networks: applications, advances and challenges, Philos. Trans. A Math. Phys. Eng. Sci. 57 (2012) 158–175.

[6] M.H. Rehmani, A.C. Viana, H. Khalife, S. Fdida, Surf: a distributed channel selection strategy for data dissemination in multi-hop cognitive radio networks, Comput. Commun. 36 (10–11) (2013) 1172–1185.

[7] M.H. Rehmani, A.C. Viana, H. Khalife, S. Fdida, A cognitive radio based internet access frame work for disaster response network deployment, in: Proceedings of the Third International Workshop on Cognitive Radio and Advanced Spectrum Management (CogART'10), in conjunction with ISABEL2010, Rome, Italy, 2010.

[8] G. Werner Allen, K. Lorincz, M. Welsh, O. Marcillo, J. Johnson, M. Ruiz, et al. Deploying a wireless sensor network on an active volcano, IEEE Internet Comput. 10 (2006) 18–25.

[9] F. Michahelles, P. Matter, A. Schmidt, B. Schiele, Applying wearable sensors to avalanche rescue, Comput. Graphics 27 (2003) 839–847.

[10] I.F. Akyildiz, E.P. Stuntebeck, Wireless underground sensor networks: research challenges, Ad Hoc Netw. 4 (2006) 669–685.

[11] P. Mitcheson, E. Yeatman, G. Rao, A. Holmes, T. Green, Energy harvesting from human and machine motion for wireless electronic devices, Proc. IEEE 96 (9) (2008) 1457–1486.

[12] M. Yuce, Introduction to wireless body area network, Wireless Body Area Networks: Technology Implementation and Applications, Jenny Stanford Publishing, Singapore, (2011).

[13] P. Bauer, M. Sichitiu, R. Istepanian, K. Premaratne, The mobile patient: wireless distributed sensor networks for patient monitoring and care, in: Proceedings 2000 IEEE EMBS International Conference on Information Technology Applications in Biomedicine, Arlington, VA, 2000, pp. 17–21.

[14] B.P.L. Lo, G.Z. Yang, Key technical challenges and current implementation of body sensor networks, in: 2nd International Workshop Wearable Implantable Body Sensor Network, Imperial College London, South Kensington, London, UK, 2005, pp. 1–5.

[15] G. Goerge, M. Kirstein, R. Erbel, Microgenerators for energy autarkic pacemakers and defibrillators, Herz 26 (1) (2001) 64–68.

[16] IEEE WPAN Task Group 4. Available from: http://www.ieee802.org/15/pub/TG4.html, 2016.

[17] IEEE 802.15 Working Group for WPAN Homepage. Available from: http://www.ieee802.org/15, 2019.

[18] T. Starner, Human-powered wearable computing, IBM Syst. J. 35 (1996) 618–629.

[19] C. Chakraborty, B. Gupta, S.K. Ghosh, A review on telemedicine-based WBAN framework for patient monitoring, Telemed, J. E Health 19 (8) (2013) 619–626.

[20] R. Yang, Q. Yong, L. Cheng, Z. Guang, W. Zhong Lin, Converting biomechanical energy into electricity by a 16 muscle-movement-driven nanogenerator, Nano Lett. 9 (2009) 1201–1205.

[21] F. Akhtar, M.H. Rehmani, Energy replenishment using renewable and traditional energy resources for sustainable wireless sensor networks: a review, Renew. Sustain. Energy Rev. 45 (2015), 769–784.

[22] C. BartonS, J. Gallaway, P. Atanassov, Enzymatic bio-fuel cells for implantable and microscale devices, Chem. Rev. 104 (2004) 4867–4886.

[23] E. Okuno, I. Caldas, C. Chow, Conservação de Energia, in: Física para ciências biológicas e biomédicas, vol. II, HARBRA, Lisboa, 1986, pp. 103–105.

[24] R.A. Bullen, T.C. Arnot, J.B. Lakeman, F.C. Walsh, Biofuel cells and their development, Biosens. Bioelectron. 21 (2006) 2015–2045.

[25] T. Starner, J.A. Paradiso, Human generated power for mobile electronics, in: Low Power Electronics Design, vol. 45, CRC Press, 2004.

[26] S. Boisseau, G. Despesse, B.A. Seddik, Electrostatic Conversion for Vibration Energy Harvesting, CEA, Grenoble, France, (2012).

[27] F. Casimiro, P.D. Gaspar, L.C.C. Gonçalves, Aplicação do princípio piezoeléctrico no desenvolvimento de pavimentos para aproveitamento energético, in: III Conferência Nacional em Mecânica de Fluidos, Termodinâmica e Energia: MEFTE—2009, Bragança, Setembro, 2009.

[28] Y.K. Ramadass, A.P. Chandrakasan, A battery-less thermoelectric energy harvesting interface circuit with 35 mV startup voltage, IEEE J. Solid State Circ. 46 (1) (2011) 333–341.

[29] R. Riemer, A. Shapiro, Biomechanical energy harvesting from human motion: theory, state of the art, design guidelines, and future directions, J. Neuroeng. Rehabil. 8 (2011) 22.

[30] K. Hernandez, R. Fernandez-Lafuente, Control of protein immobilization: coupling immobilization and site-directed mutagenesis to improve biocatalyst or biosensor performance, Enzyme Microb. Technol. 48 (2) (2011) 107–122.

[31] S.D. Minteer, B.Y. Liaw, M.J. Cooney, Enzyme-based biofuel cells, Curr. Opin. Biotechnol. 18 (2007) 228–234.

[32] K. Scott, E.H. Yu, M.M. Ghangrekar, B. Erable, N.M. Duteanu, Biological and microbial fuel cells, Compr. Renew. Energy 4 (2012) 277–300.

[33] E.H. Yu, S. Keith, Enzymatic biofuel cells—fabrication of enzyme electrodes, Energies 3 (2010) 23–42.

[34] D. Morton, Human Locomotion and Body Form, The Williams & Wilkins Co., Baltimore, (1952).

[35] M. Ashraf, N. Masoumi, A thermal energy harvesting power supply with an internal startup circuit for pacemakers, in: IEEE Transactions on Very Large Scale Integration (VLSI) Systems, IEEE, 2015

[36] X. Wei, J. Liu, Power sources and electrical recharging strategies for implantable medical devices, Front. Energy Power Eng. China 2 (2008) 1–13.

[37] D.A. Winter, Biomechanics and Motor Control of Human Movement, 3 ed., John Wiley and Sons, Hoboken, NJ, (2005).

[38] C-M. Kyung, Nano Devices and Circuit Techniques for Low-Energy Applications and Energy Harvesting, Springer, The Netherlands, 2016.

[39] Energy harvesting from the human body and powering up implant devices, KAIST Research Series, 2016.

[40] A.B. Amar, A.B. Kouki, H. Cao, Power approaches for implantable medical devices, Sensors 15 (2015) 28889–28914.

[41] A. Telba, W.G. Ali, Modeling and simulation of piezoelectric energy harvesting, in: Proceedings of the World Congress on Engineering 2012.

[42] Publitek Marketing Communications. Available from: http://www.digikey.com/en/articles/techzone/2014/nov/piezoelectric-energy-harvesting-at-the-heart-of-the-matter, 2014.

[43] Available from: http://www2.imec.be/be_en/press/imec-news/archive-2007/imec-realizes-wireless-eeg-system-powered-by-body-heat.html, 2007.

[44] A. Tabesh, L.G. Fréchette, A low-power stand-alone adaptive circuit for harvesting energy from a piezoelectric micropower generator, IEEE Trans. Indust. Electron. 57 (3) (2010) 840–849.

[45] R. Wylie, Your body, the battery: powering gadgets from human "biofue", 2015.

[46] P.P. Mercier, A.C. Lysaght, S. Bandyopadhyay, A.P. Chandrakasan, K. Mstankovic, Energy extraction from the biologic battery in the human ear, Nat. Biotechnol. 30 (2012) 1240–1243.

[47] M. Yip, R. Jin, H.H. Nakajima, K.M. Stankovic, A.P. Chandrakasan, Fully-implantable cochlear implant SoC with piezoelectric middle-ear sensor and arbitrary waveform neural stimulation, IEEE J. Solid State Circ. 50 (2015) 214–229.

[48] G. Terlecka, A. Vilumsone, J. Blums, I. Gornev, The structure of the electromechanical converter and its integration in apparel, Sci. J. Riga Tech. Univ. Mater. Sci. Textile Cloth. Technol. 6 (2011) 123.

[49] A. Delnavaz, J. Voix, Flexible Piezoelectric Energy Harvesting from Jaw Movements, IOP Science, Bristol, (2014).

[50] Available from: http://www.yankodesign.com/2011/10/10/human-hampster-wheel-for-energy.

[51] H. Huang, X. Li, Y. Sun, A triboelectric motion sensor in wearable body sensor network for human activity recognition, Conf. Proc. IEEE Eng. Med. Biol. Soc. 2016 (2016) 4889–4892.

[52] S. Almouahed, C. Hamitouche, P. Poignet, E. Stindel, Battery-free force sensor for instrumented knee implant, in: 2017 IEEE Healthcare Innovations and Point of Care Technologies (HI-POCT), IEEE, Bethesda, MD, USA, pp. 1–4.

[53] M.A. Halim, H.O. Cho, J.Y. Park, A handy motion driven, frequency up-converting piezoelectric energy harvester using flexible base for wearable sensors applications, in: IEEE SENSORS, IEEE, Busan, South Korea, 2015, pp. 1–4.

[54] H. Fu, K. Cao, R. Xu, M.A. Bhouri, R. Martínez-Botas, S.-G. Kim, E.M. Yeatman, Footstep energy harvesting using heel strike induced airflow for human activity sensing, in: IEEE 13th International Conference on Wearable and Implantable Body Sensor Networks (BSN), 2016, IEEE, San Francisco, CA, USA, pp. 124–129.

[55] D. Yun, J. Park, K.-S. Yun, Highly stretchable energy harvester using piezoelectric helical structure for wearable applications, Electron. Lett. 51 (3) 284–285.

[56] D.-S. Kwon, H.-J. Ko, J. Kim, Piezoelectric and electromagnetic hybrid energy harvester using two cantilevers for frequency up-conversion, in: 2017 IEEE 30th International Conference on Micro Electro Mechanical Systems (MEMS), IEEE, Las Vegas, NV, USA, 2017.

[57] J.H. Kim, S.J. Hwang, Y. Song, C.H. Yang, M.S. Woo, K.J. Song, T.H. Sung, Designing a piezoelectric energy harvester using clicking mechanism, in: 5th International Conference on Renewable Energy Resources and Application, IEEE, Birmingham, UK, 2016.

[58] S. Kumari, S. Sahu, B. Gupta, Efficient SSHI circuit for piezoelectric energy harvester uses one shot pulse boost converter, Analog Integr. Circuits Signal Process. 97 (3) (2018) 545–555.

[59] N. Dey, A.S. Ashour, F. Shi, S.J. Fong, R.S. Sherratt, Developing residential wireless sensor networks for ECG healthcare monitoring, IEEE Trans. Consum. Electron. 63 (4) (2017) 442–449.

[60] G. Elhayatmy, N. Dey, A.S. Ashour, Internet of things based wireless body area network in healthcare, in: Internet of Things and Big Data Analytics toward Next-Generation Intelligence, Springer, Cham, 2018, pp. 3–20.

[61] N. Dey, V. Bhateja, A.E. Hassanien, Medical Imaging in Clinical Applications, Springer International Publishing, Switzerland, (2016).

[62] C. Bhatt, N. Dey, A.S. Ashour, A.S. (Eds.), Internet of Things and Big Data Technologies for Next Generation Healthcare, Springer, Cham, 2017.

[63] N. Dey, A.S. Ashour, F. Shi, S.J. Fong, J.M.R. Tavares, Medical cyber-physical systems: a survey, J. Med. Syst. 42 (4) (2018) 74.

BER performance comparison of EEG healthcare system using 8-pixel, 16-pixel OLED screen, and DSLR camera

Geetika Aggarwal[a], **Xuewu Dai**[a], **Reza Saatchi**[b], **Richard Binns**[a]

[a]Northumbria University, Newcastle upon Tyne, United Kingdom; [b]Sheffield Hallam University, Sheffield, United Kingdom

1 Introduction

To extend existing communication services in terms of capacity, connectivity, and to provide low power consumption, fifth generation (5G) communication is the upcoming direction. Nowadays, the basic principle to transmit data wirelessly is through protocols of radio frequency (RF). However, the RF electromagnetic spectrum is populous, hence is struggling to reach the increasing demand of end users [1–4]. The electromagnetic spectrum range of RF band is licensed and regulated. Hence, to meet the growing demand of data communication, substitute is optical wireless communication (OWC) as it has unique advantages over RF, the foremost being the huge unlicensed bandwidth spectrum, a major requirement for data communication. The VLC, member of OWC, encompasses light emitting diode (LEDs) being transmitter and photodiode (PD) as receiver but the utilization of camera or image sensor as receiver in communication systems; a supplement of VLC is known as optical camera communication (OCC) [5–10]. The advantage of OCC is that it does not need hardware modifications; combination of VLC and OCC is known as Visible Light Optical Camera Communication [10,12].

In 1924, Hans Berger invented electroencephalography (EEG), which is a noninvasive mechanism to gather the brain information using scalp electrodes. The usual EEG recording is tedious and time-consuming with several scalp electrodes connected through wires to the data analysis screen. The wireless wearable EEG headsets such as Neurosky, Imec, emotiv, etc., shown in Fig. 6.2 are the consequences of advancement in RF-based wireless technology. These headsets are used in brain monitoring for specific applications like sleeping disorders, cognitive analysis, and many more depending upon the specific applications and the number of channels

Smart Biosensors in Medical Care. http://dx.doi.org/10.1016/B978-0-12-820781-9.00006-1

required for EEG applications [13,14]. The wireless technology-based products deployed in healthcare in most of the applications such as to record the heart activity, electrocardiogram (ECG) which is noninvasive procedure. In [15] wireless monitoring system for EEG using Zigbee is proposed due to low cost and low power consumption. In [16], authors presented a multi-resolution wavelet transform-based system for detection "P," "Q," "R," "S," "T" peaks complex from original ECG signal. "R-R" time lapse is an important minutia of the ECG signal that corresponds to the heartbeat of the concerned person. Abrupt increase in height of the "R" wave or changes in the measurement of the "R-R" denote various anomalies of human heart thus providing the useful of human heart through ECG. However, in recent years, due to the advancement in wireless technology the quality of healthcare has significantly increased at hospitals and healthcare environments. Therefore, in RF sensitive areas such as intensive care units (ICUs), because of RF, electromagnetic interferences (EMIs) to other medical electronic devices have crucial impact on both humans and medical devices. Hence, this chapter displays the lab work for EEG system using VL-OCC, where the experiments were conducted in LOS using 8-pixel and 16-pixel organic light emitting diode (OLED) screen as transmitter and DSLR camera as receiver. The significant contributions of this chapter are listed as follows:

- The results of the proposed system were analyzed and compared in terms of frames, bits transmitted, and length of video using DSLR camera.
- Experimentally demonstrated the trade-off between the bit rate and BER depending upon the pixel size.

The remaining section is categorized into following sections: Section 2 focuses on VLC in healthcare, briefly listing the related work, advantages of VLC over RF, and finally the VL-OCC definition and its scope. Section 3 illustrates the proposed system modeling followed by Section 4 with experiments and Section 5 with results. Finally, this chapter is concluded in Section 6.

2 VLC in healthcare

The application of EEG in hospitals or in healthcare involving wet electrodes or sensors to measure the scalp activity is done using RF technology that might interfere with the medical equipment in addition to the patient's health due to EMI especially in RF restricted zone areas such as ICU. Furthermore, RF spectrum is congested and crowded; hence, in view of issues related to RF in healthcare, an alternative source of communication is essential. VLC, operating within the range of 400 –780 nm is the optimum choice in healthcare over RF due to following:

A. Safe for the human health

The use of the LED light as a transporter for the information empowers VLC to be totally suitable for the human well-being. The RF negative impact on the human well-being is increasingly inclined to RF prohibited regions explicitly hospitals. Being protected to human well-being empowers the likelihood of VLC in information correspondence in medicinal services or healthcare [17].

B. Security

On the contrary to RF waves, the light cannot enter through dividers, giving VLC high security against eavesdropping. In VLC, one can fundamentally observe the information and guarantee the security of the information basically by shutting the entryway. This makes VLC to be reasonable in military applications or in regions of high security, for example, medical clinic where the patient information should have been kept confidential [8–9,17].

C. Low-cost implementation

In contrast to RF that utilizes a directed band, VLC utilizes the light for data communication, which is in an unlicensed district of the electromagnetic range. Since no expense for a permit is inferred, the cost of implementation is remarkably decreased. Another advantage of VLC that reduces the cost of implementation is pervasive nature. VLC depends on existing frameworks that are as of now acknowledged and broad over the world, on account of the various favorable circumstances offered by LED lighting sources. This component will make the execution of VLC easy and simple, without requiring urge adjustments on the current infrastructure. The third characteristic which empowers VLC to decrease the cost of implementation is its diminished multifaceted nature. VLC essentially utilizes LED producers and photo-diode beneficiaries, segments which are reasonable and inexpensive in context to RF infrastructure [8,17].

D. Green wireless communication technology

While the Earth's populace is expanding and the human culture is building up, the common asset utilization and the atmosphere weakening are additionally expanding. Ozone depleting substance outflows have achieved disturbing dimensions that are delivering significant atmosphere changes that influence the entire environment [9]. The normal asset utilization and the contamination can be altogether diminished by diminishing the vitality utilization. Counterfeit lighting, usually given by electric lights, speaks to a noteworthy percent of the vitality utilization. Around the world, roughly 19% of power is utilized for lighting, while power speaks to 16% of the complete vitality created [9,18]. Other than the upper referenced points of interest, VLC is additionally a green remote correspondence innovation. VLC is green technology since it does not utilize extra power for the correspondence. A similar light which is utilized for lightning is utilized for conveying the information. Another significant favorable position of VLC is the utilization of LEDs which gives generous vitality reserve funds, diminishing the CO_2 emanations [9,18,19].

The aforementioned points of VLC over RF make VLC the best candidate for healthcare applications specifically hospitals by combining the idea of lightning and communication together for efficient wireless data services with no RF radiation.

2.1 VLC system

The LED lightning is replacing the existing incandescent lighting systems due to low cost, longer life span, low power consumption, and dual functionality of lightning and data communication. VLC employs LED at transmitter section and photodiode at receiver section, where transmitter is electro-optical device converting electrical

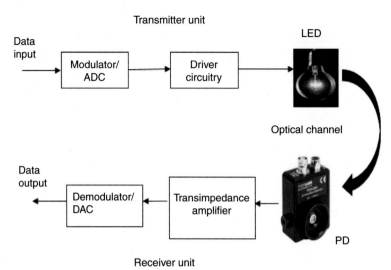

FIGURE 6.1 Basic VLC system block diagram.

signal to optical and photodiode is optical-electronic device thus converting optical signal to electrical [7,20]. The basic VLC system diagram is shown in Fig. 6.1.

The data input undergoes modulation before being transmitted using LED over the optical channel in VLC system. The transmitted data are in the form of light intensity captured by photodiode to receiver unit, followed by demodulation and offline processing to get the output data. Depending upon the intensity of the light signal received and the gain of the amplifier specified, transimpedance amplifier (TIA) amplifies the received signal by photodiode. The output data analysed further both offline and online to check the system reliability and BER performance.

2.2 Receivers in VLC system

VLC systems deploy two receivers categorized as:

2.2.1 Photodetector

The commonly used photodiodes in VLC systems are PIN PD and avalanche PD (APD), the latter having a higher gain. The APD has a drawback of extra noise generated affecting the performance of the system. On the other hand, PDs are predominantly used in VLC systems because of low cost and high temperature tolerance [3–7].

2.2.2 Imaging sensor or camera—also called a camera sensor

Cameras are pervasive thus extensively been deployed for both data communication and photography. The computer vision and image processing has made the communication easier with the help of quick response (QR) codes. These QR codes prompt in camera communication applications, using 2D barcodes [11,12].

2.3   Related work on healthcare system using VLC

The authors in [21] presented a VLC system based on time hopping (TH) using LED/PD pair for ECG, the advantage of the system being that it could be used in RF restricted areas and in emergency due to security and free from EMI. In [22] authors presented a simple VLC-based indoor monitoring system deploying OOK modulation scheme. The LED/PD was used for the transmission and reception of data through sensors thus providing an efficient communication system free from EMI and safe to both humans and medical equipment. In [23] authors demonstrated a dust monitoring system using VLC deploying wavelength division multiplexing (WDM) through RGB LEDs. The incoming light signals were differentiated using a color sensor and then received and processed by a microcontroller thus enabling the medical team to monitor the indoor air quality and improve the environmental conditions of hospitals. The authors in [24] proposed a VLC system for EEG signal transmission using three RGB LEDs as transmitter and three photodiodes as receiver. In [20] the authors proposed a VLC-based LabVIEW system for single channel EEG transmission using LED/PD pair. Experiments performed in LOS with OOK-NRZ modulation resulted in improvement in SNR to 12.29 dB at 1 m. The authors in [21] proposed a VLC system for ECG, which is noninvasive process to record the electrical activity of heart deploying LED/PD pair using TH scheme. The authors in [22] presented a low data rate static patient monitoring system in hospitals based on VLC using OOK modulation scheme deploying LED/PD pair. The proposed system was EMI free in comparison to RF protocols. As VLC systems use LED and PD as transmitter and receiver, respectively, the authors in [25] proposed a medical portable device to download and access the healthcare information connected to optical sensor using high brightness LED (HB-LED) and high-speed photodetector. Robotic and artificial intelligence has been used widely to ease the workload of humans, hence in [26–28] authors are working on VLC system of localization for robots to be used in healthcare by employing robot sensors in the specific area to be targeted so that communication is not hampered.

2.4   VLC in healthcare—state of the art

The essential favorable position that VLC conveys to social insurance applications is the end of obstruction that RF-based medicinal gadgets endure. Different advantages in VLC incorporate lower cost than RF interchanges frameworks; better correspondence among hardware, clinical staff and patients, and checking stations; more noteworthy security of light flag contrasted and RF flag; and complete well-being amid stretched out presentation to patients [29]. A lot of work is going on in world on VLC especially for healthcare applications, some of them are listed later.

2.4.1   Pukyong National University, South Korea

Wearable EEGs can monitor brain–computer interfaces, fatigue, and emotional and mental states, but the RF wireless radiation posed long-term risks to the patient's brain. The Pukyong National University, South Korea, team avoided this hazard by integrating VLC into a wearable EEG device *equipped* with electrodes to gather brain activity data, and a microcontroller as an analog data filter and analog-to-digital converter. The researchers are considering this to be an alternative to Bluetooth technology due to VLC being safe for human well-being [30].

2.4.2 National Taiwan University, Graduate Institute of Photonics and Optoelectronics

Scientists and researchers at National Taiwan University have developed a violet laser diode as an alternative to the LEDs typically used in VLC systems for indoor lighting. The researchers refined the chemistry and adjusted the phosphor thickness of their violet laser's diffuser plats to achieve a data rate capacity as high as 12 gigabits per second at distances over 7 m. If greater volumes of data could be transmitted by VLC systems, it would also make the technology more attractive to healthcare providers [29].

2.4.3 Moriya Research Laboratory, NTT Communication Science Laboratories, Kyoto, Japan

Scientists and *researchers* at Kyoto, Japan, have built up a VLC framework that utilizes a camcorder and an occasion timing encoding method to screen numerous pulses. The LED on the wearable heart screen flashes a sign comparing to every heartbeat that is caught by the camera. The encoding highlight records and breaks down the heartbeat to distinguish a hazard and ready parental figures or people on call [29].

The earlier *listed* some of the research work of VLC in healthcare makes VLC a best alternative in healthcare replacing RF. Furthermore, the camera being receiver with minimum hardware requirement opens the door for VL-OCC in healthcare.

2.5   Visible light optical camera communication

VL-OCC not only aids in cost reduction of the communication systems due to minimal hardware *requirement* but also reduces the extra noise and blurring by focusing the camera at the target [31–37]. Block diagram of VL-OCC is listed in Fig. 6.2.

In Fig. 6.2, the data input is modulated and then transmitted to light for transmission through the optical channel, where the camera takes the video/image. This video/image is then offline processed using image processing and computer vision to extract the data bits, compare them with the input data and calculate BER.

---

3   Methodology of EEG VL-OCC healthcare

The methodology is explained using Fig. 6.3, where, b_n denotes the transmitted data bits and $N = R_1 \times C_1$ is the matrix formed by rows and columns on OLED screen. The signal bits are sent to the ARM processor, and then to the transmitter, which is OLED screen followed by conversion, serial to parallel (*S/P*), hence signal becomes $s^{(t)} = [r_1, c_1]$. Depending on pixel size, D given, data bits transmitted will be:

$$\left(S_{z_{row}}/D\right) \times \left(S_{z_{column}}/DM\right) = R_1 \times C_1 \tag{1}$$

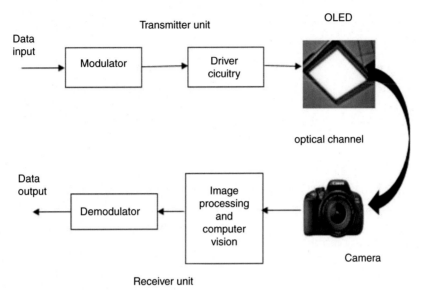

FIGURE 6.2 Block diagram of VL-OCC.

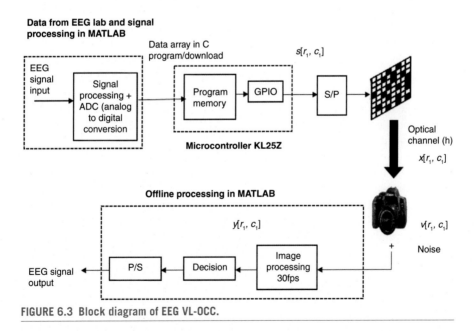

FIGURE 6.3 Block diagram of EEG VL-OCC.

where $S_{z_{row}}$ and $S_{z_{column}}$ denote respective rows and columns. The transmission through OLED screen, the optical signal can be presented by:

$$x^{(t)}[r_1, c_1] = (s * h)^{(t)}[r_1, c_1] \qquad (2)$$

In Eq. (2), h states the impulse response; therefore, received signal, $y^{(t)}[r_1,c_1]$, becomes:

$$y^{(t)}[r_1,c_1]=(x)^{(t)}[r_1,c_1]+(v)^{(t)}[r_1,c_1] \tag{3}$$

where $(v)^{(t)}[r_1,c_1]$ is the dominant White Gaussian noise (WGN) which is ambient light in this research work. The noise is reduced to zero or minimum by using the region of interest during the capture of video or images.

4 Experimental work and hardware description

This section lists the experimental work performed using 8-pixel and 16-pixel screen of OLED being transmitter and DSLR camera being receiver and hardware description. EEG signal is a mixture of several frequencies ranging from 4 Hz to 50 Hz and the amplitude of EEG signal in millivolts or microvolts because the EEG signal is obtained through sensors beneath the scalp [38]. The EEG signal for experimental work was obtained from EEG toolbox [39] and is plotted in MATLAB shown in Fig. 6.4A.

The EEG signal toolbox provides researchers with several EEG signals with several channels for research purpose. In this research single channel EEG signal was used. EEG toolbox also provides the facility to filter, amplify, and compress the signal before downloading in MATLAB [39], hence in this chapter and research the obtained EEG signal was compressed to enhance the data bit transmission rate. As the intensity of the signal in VL-OCC system cannot be negative, the EEG signal is normalized between 0 and 1, shown in Fig. 6.4B. For the analog to digital conversion, a 16-bit ADC was chosen to allow for higher resolution and low quantization error. The bits obtained were up sampled to avoid aliasing and then uploaded to microcontroller through MBED using C language. Table 6.1 lists the equipment used in experimental work.

The microcontroller of KL series, FRDM-KL25Z, was made the choice due to readily available, cost-effective, and low voltage requirement. DD-160128FC OLED [40] is a transmitter which is a screen of OLED and has 3.6 g of weight. The driver circuit for OLED requires 2.8 V, hence switching board made of potential dividers on PCB using KICAD software was designed. The experiments were performed in LOS at several distances; camera was used at receiver section to capture the incoming signal in the form of video and then processed further for offline processing in MATLAB to calculate BER. Figs. 6.5 and 6.6 show the hardware setup and the information bits transmitted through OLED screen, respectively, which are then captured by camera for offline processing.

After signal processing, the data bits are sent to microprocessor (1) using MBED software through USB cable, which are then transferred to the transmitter (3) with the help of PCB switching board (2) and the power supply is maintained using a voltage regulator (4). As per the data sheets the OLED screen made up of rows and columns varies with the size of pixel. For instance, if the pixel size is small then there is increase in information bits that can be sent in single frame and vice versa. For

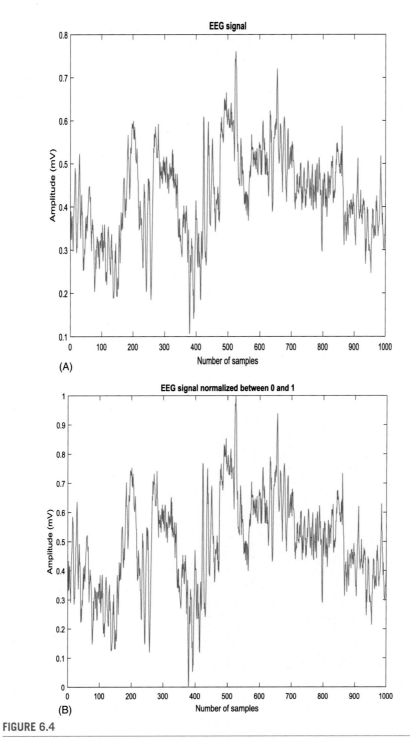

FIGURE 6.4

(A) EEG signal from EEG lab, (B) EEG signal normalized between 0 and 1.

Table 6.1 Equipment used.

Equipment	Model
Optical transmitter	OLED DD-160128FC-1A
Embedded system	FRDM-KL-25Z
Camera	Canon DSLR
Voltage requirement	2.8V for logic
Programming software	KICAD, C, MATLAB

the 8-pixel OLED screen, data sheets of OLED screen depict that the transmitted is dissected into 16 rows and 20 columns, thus total of 320 bits can be transmitted per frame. Similarly, for 16-pixel OLED screen is split into 8 rows and 10 columns thus aggregate of 80 bits are transmitted through single frame. The camera parameters or specifications are listed in Table 6.2.

After the transmission of bits, DSLR camera was used to capture the video and process it later for image processing in MATLAB to calculate BER. The length of the video was dependent on the bit rate and the number of total bits can be transmitted per frame. For instance, to transmit 2 kbps bit rate with the capacity 320 bits through a single frame of 8-pixel OLED screen, the length of video required would be:

$$\text{Length of video required} = \frac{\text{Total bit rate}}{\text{Number of bits can be transmitted per frame}} \tag{4}$$

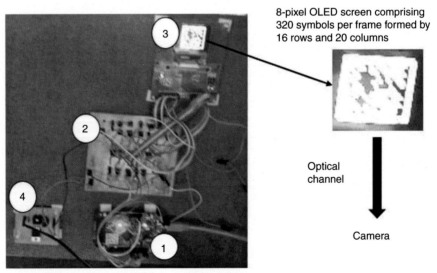

8-pixel OLED screen comprising 320 symbols per frame formed by 16 rows and 20 columns

Optical channel

Camera

FIGURE 6.5 Hardware description of 8-pixel OLED screen.

1, Microprocessor, FRDM, KL-25Z; *2*, PCB switching board; *3*, transmitter—8 pixel; *4*, voltage regulator.

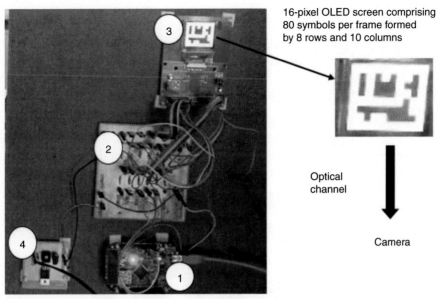

16-pixel OLED screen comprising 80 symbols per frame formed by 8 rows and 10 columns

Optical channel

Camera

FIGURE 6.6 Hardware description of 16-pixel OLED screen.

1, Microprocessor, FRDM, KL-25Z; *2*, PCB switching board; *3*, transmitter—16 pixel; *4*, voltage regulator.

$$\text{Length of video required} = \frac{2.8 \text{ kbps}}{320} 8.75 \text{ s} \tag{5}$$

And for 16-pixel OLED screen with 80 data bits capacity per frame, length of video would be:

$$\text{Length of video required} = \frac{2.8 \text{ kbps}}{80} = 35 \text{ s} \tag{6}$$

The image processing of the video captured by a camera is done in MATLAB to measure the BER performance of 8-pixel and 16-pixel OLED screen. The next section lists the results obtained and image processing in MATLAB.

Table 6.2 Camera specifications.

Type of camera	DSLR
Camera resolution	18 MP
Camera frame rate	30 fps
Autofocus	Manual
Data rate achieved	2.8 kbps
Exposure mode	Rolling shutter

5 Results and discussion

Throughout the lab work, several experiments were conducted to analyze the BER performance. Figs. 6.7 and 6.8 depict the 8-pixel and 16-pixel image processing. For instance, if the bit transmitted is 1 then in the received image it is represented by white section and if the bit transmitted is 0 then in the received image it is represented as 0.

In order to maintain the synchronization between the transmitted and camera video capture, a preamble of 400 bits of 1s was initially sent to indicate the start of the data bits and the end of video capture. The advantage of deploying camera as

FIGURE 6.7 Pixel size 8, 320 bits per frame.

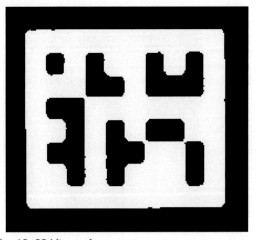

FIGURE 6.8 Pixel size 16, 80 bits per frame.

receivers is that due to focusing properties and inbuilt filter there is no need of hardware or infrastructure modifications, hence by using the selective capture technique of camera and focusing the camera blurring of the images was mitigated during the experimental work. It is clear from Figs. 6.7 and 6.8 that due to small pixel size, more bits were transmitted through the 8-pixel OLED screen in comparision to 16-pixel OLED screen; however, the BER increases considerably for the 8 pixel in context to 16 pixel as shown in Fig. 6.9.

In image processing the images extracted from video capture during the experimental work are compared with the transmitted images or bits to calculate BER by:

$$BER = \frac{\text{Transmitted bits}}{\text{Length of bits}} \qquad (7)$$

The results depict that the error-free transmission up to a distance of 1.75 m or 175 cm was achieved with 8-pixel OLED screen in comparison to 16-pixel OLED screen, where error-free transmision link distance is 2.25 m or 225 cm. Hence, there

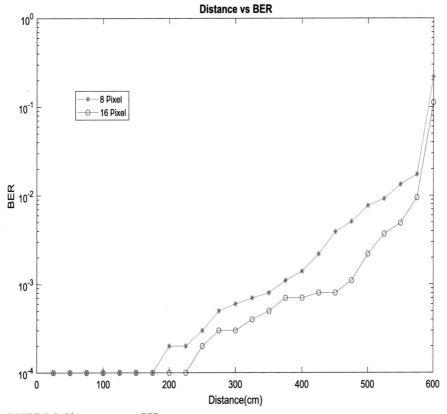

FIGURE 6.9 Distance versus BER.

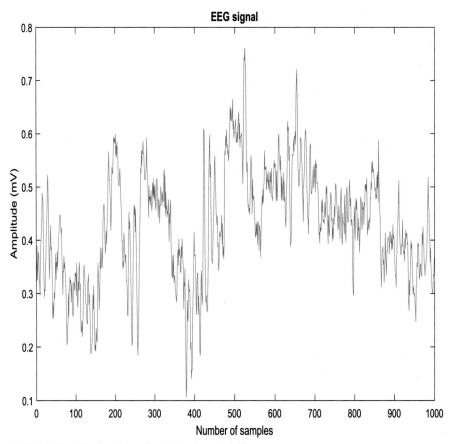

FIGURE 6.10 EEG signal received after demodulation.

is a trade-off between the pixel size and the total number of transmitted data bits, for instance, more data bits can be transmitted with smaller pixel size but BER increases significantly for the same link distance.

The received EEG signal shown in Fig. 6.10 is obtained after offline and image processing, demodulation, and digital to analog conversion. The EEG signal received as shown in Fig. 6.10 is like the EEG signal transmitted in Fig. 6.4A, therefore indicating the efficient transmission of EEG using transmitter to be visible light OLED and receiver to be camera.

6 Conclusions

This chapter demonstrated the BER performance of the VL-OCC EEG healthcare system; the transmitter was OLED screen with varying pixel sizes 8 and 16, the DSLR camera being the receiver. The use of camera at receiver section does not need hardware modification due to inbuilt filters and focusing properties in context to conventional VLC systems. Furthermore, the proposed VL-OCC system is best suited in healthcare as free from harmful radiations such as EMI in comparison to RF protocols used in healthcare, which suffers from EMI and has adverse effect on patient's health in RF restricted areas. The BER performance comparison among 8-pixel and 16-pixel transmitter shows that there is a trade-off between the pixel size and the number of bits transmitted. Though more data bits can be transmitted through the 8-pixel OLED screen, the BER increases significantly after 1.75 m or 175 cm in comparison to 16-pixel OLED screen where the error-free transmission was achieved until 2.25 m or 225 cm. Furthermore, due to system on chip prototype, low cost, low power consumption, and EMI free, the proposed could be used for wireless remote monitoring in both hospitals and BCI in context to EEG applications.

References

[1] P. Pirinen, A brief overview of 5G research activities, in: Proceeding of the International Conference on 5G Ubiquitous Connectivity (5GU), IEEE, Akaslompolo, Finland, 2014, pp. 17–22.

[2] M. Shafi, et al. 5G: a tutorial overview of standards, trials, challenges, deployment, and practice, IEEE J. Sel. Areas Commun. 35 (6) (2017) 1201–1221.

[3] M. Uysal, H. Nouri, Optical wireless communications—an emerging technology, in: Proceeding of the International Conference on Transparent Optical Networks (ICTON), IEEE, Graz, Austria, 2014, pp. 1–7.

[4] M. Zaman, et al. A comparative survey of optical wireless technologies: architectures and applications, IEEE Access 6 (2018) 9819–9840.

[5] A.C. Boucouvalas, P. Chatzimisios, Z. Ghassemlooy, M. Uysal, K. Yiannopoulos, Standards for indoor optical wireless communications, IEEE Commun. Mag. 53 (3) (2015) 24–31.

[6] Z. Ghassemlooy, S. Arnon, M. Uysal, Z. Xu, J. Cheng, Emerging optical wireless communications-advances and challenges, IEEE J. Sel. Areas Commun. 33 (9) (2015) 1738–1749.

[7] Z. Ghassemlooy, L.N. Alves, S. Zvanovec, M.A. Khalighi, Visible Light Communications: Theory and Applications, CRC Press, Boca Raton, FL, (2017).

[8] H. Chen, H.P.A. van den Boom, E. Tangdiongga, T. Koonen, 30-Gb/s bidirectional transparent optical transmission with an MMF access and an indoor optical wireless link, IEEE Photon. Technol. Lett. 24 (7) (2012) 572–574.

[9] D.K. Borah, A.C. Boucouvalas, C.C. Davis, S. Hranilovic, K. Yiannopoulos, A review of communication-oriented optical wireless systems, EURASIP J. Wireless Commun. Netw. 2012 (2012) 91.

[10] R.D. Roberts, Undersampled frequency shift ON-OFF keying (UFSOOK) for camera communications (CamCom), in: Wireless and Optical Communication Conference, IEEE, Chongqing, China, 2013, pp. 645–648.

[11] N. Rajagopal, P. Lazik, A. Rowe, Hybrid visible light communication for cameras and low-power embedded devices, in: Proceeding of the ACM MobiCom Workshop on Visible Light Communication Systems, ACM, New York, 2014, pp. 33–38.

[12] Z. Ghassemlooy, P. Luo, S. Zvanovec, Optical camera communications, in: Optical Wireless Communications, Springer, Cham, Switzerland, 2016, pp. 547–568.

[13] What to Know about EEG Tests. Available from: https://www.medicalnewstoday.com/articles/325191.php.

[14] V. Mihajlovic, B. Grundlehner, R. Vullers, J. Penders, Wearable, wireless EEG solutions in daily life applications: what we are missing? IEEE J. Biomed. Health Informatics 25 (1) (2015) 615–618.

[15] N. Dey, A.S. Ashour, F. Shi, S.J. Fong, R.S. Sherratt, Developing residential wireless sensor networks for ECG healthcare monitoring, IEEE Transact. Consum. Electron. 63 (4) (2017) 442–449.

[16] S. Mukhopadhyay, S. Biswas, A.B. Roy, N. Dey, Wavelet based QRS complex detection of ECG, International Journal of Engineering Research and Applications (IJERA) 2(3),2012, 2361-2365.

[17] M.K. Hasan, M. Shahjalal, M.Z. Chowdhury, Y.M. Jang, Access point selection in hybrid OCC/RF eHealth architecture for real-time remote patient monitoring, in: 2018 International Conference on Information and Communication Technology Convergence (ICTC), IEEE, Jeju, 2018, pp. 716–719, doi:10.1109/ICTC.2018.8539582.

[18] S. Dimitrov, H. Haas, Information rate of OFDM-based optical wireless communication systems with nonlinear distortion, J. Lightw. Technol. 31 (6) (2013) 918–929.

[19] H. Elgala, R. Mesleh, H. Haas, Indoor optical wireless communication: potential and state-of-the-art, IEEE Commun. Mag. 49 (9) (2011) 56–62.

[20] G. Aggarwal, X. Dai, R. Binns, R. Saatchi, Experimental demonstration of EEG signal transmission using VLC deploying LabView, in: 2018 3rd International Conference and Workshops on Recent Advances and Innovations in Engineering (ICRAIE), IEEE, Jaipur, India, 2018, pp. 1–6.

[21] J. An, W. Chung, A novel indoor healthcare with time hopping-based visible light communication, in: 2016 IEEE 3rd World Forum on Internet of Things (WF-IoT), IEEE, Reston, VA, 2016, pp. 19–23.

[22] W.A. Cahyadi, T. Jeong, Y. Kim, Y. Chung, T. Adiono, Patient monitoring using visible light uplink data transmission, in: 2015 International Symposium on Intelligent Signal Processing and Communication Systems (ISPACS), IEEE, Nusa Dua, 2015, pp. 431–434.

[23] J. An, W. Chung, Wavelength-division multiplexing optical transmission for EMI-free indoor fine particulate matter monitoring, IEEE Access 6 (2018) 74885–74894.

[24] D.R. Dhatchayeny, A. Sewaiwar, S.V. Tiwari, Y.H. Chung, Experimental biomedical EEG signal transmission using VLC, IEEE Sens. J. 15 (10) (2015) 5386–5387.

[25] X.-W. Ng, W.-Y. Chung, VLC-based medical healthcare information system. Biomed. Eng. Appl. Basis Commun. 24 (2012), doi:10.4015/S1016237212500123.

[26] F. Dellaert, D. Fox, W. Burgard, S. Thrun, Monte Carlo localization for mobile robots, in: Proceeding of the IEEE International Conference on Robotics & Automation (ICRA), IEEE, Detroit, MI, USA, 1998.

[27] R. Ueda, T. Arai, K. Asanuma, K. Umeda, H. Osumi, Recovery methods for fatal estimation errors on Monte Carlo localization, J. Robot. Soc. Japan 23 (4) (2005) 466–473.

[28] R. Murai et al., A novel visible light communication system for enhanced control of autonomous delivery robots in a hospital, in: 2012 IEEE/SICE International Symposium on System Integration (SII), IEEE, Fukuoka, 2012, pp. 510–516.

[29] Impact of Visible Light Communication in Healthcare. Available from: https://aabme. asme.org/posts/impact-of-visible-light-communication-technology-in-health-care.

[30] T. Zhou, X. Lee, L. Chen, Temperature monitoring system based on Hadoop and VLC, Procedia Comput. Sci. 131 (2018) 1346–1354.

[31] N. Iizuka, OCC proposal of scope of standardization and applications, IEEE 802.15 SG7a standardization documents, 2014.

[32] K. Kuraki, S. Nakagata, R. Tanaka, T. Anan, Data transfer technology to enable communication between displays and smart devices, FUJITSU Sci. Tech. J. 50 (1) (2014) 40–45.

[33] Y.-S. Kuo, P. Pannuto, K.-J. Hsiao, P. Dutta, Luxapose: indoor positioning with mobile phones and visible light, in: Proceeding of the Annual International Conference on Mobile Computing and Networking, ACM, Maui, HI, USA, 2014, pp. 7–11.

[34] W. Hu, H. Gu, Q. Pu, LightSync: unsynchronized visual communication over screen-camera links, in: Proceedings of the 19th Annual International Conference on Mobile Computing & Networking, ACM, New York, 2013, pp. 15–26.

[35] G. Aggarwal, X. Dai, R. Saatchi, R. Binns, A. Sikandar, Experimental demonstration of single-channel EEG signal using 32×32 pixel OLED screen and camera, Electronics 8 (2019) 734.

[36] H.-Y. Lee, Unsynchronized visible light communications using rolling shutter camera: implementation and evaluation. M.S. thesis, Department of Computer Science and Informatics Engineering, College of Electrical Engineering & Computer Science, National Taiwan University, 2014, pp. 1–55.

[37] T. Nguyen, N.T. Le, Y.M. Jang, Asynchronous scheme for unidirectional optical camera communications (OCC), in: International Conference on Ubiquitous and Future Networks (ICUFN), IEEE, Shanghai, China, 2014, pp. 48–51.

[38] S. Chatterjee, A. Miller, Biomedical Instrumentation Systems, Delmar Cengage Learning, Clifton Park, NY, 2010.

[39] Etoolbox, Bioelectromagnetism MATLAB Toolbox. Available from: http://eeg.sourceforge.net.

[40] DD-160128FC-1A DENSITRON, Graphic OLED, 160×128, RGB, 2.8V, Parallel, Serial, 35.8mm $\times$ 30.8mm, −20 °C, Farnell UK. Available from: https://uk.farnell.com/densitron/dd-160128fc-1a/display-oled-rgb-160x128/dp/1498857.

FBCC protocol for HBAN network for smart e-health application

7

Ankur Dumka[a,b], **Alaknanda Ashok**[a,b]

[a]Graphic Era Deemed to be University, Dehradun, India;
[b]G.B. Pant University of Agriculture and Technology, Pantnagar, India

1 Introduction

The modern technologies like wireless communication, advanced networks, nano-technology, creature science, medical science, etc., have changed the accomplishment and trends of modern technology. Wireless body area network (WBAN) or body area network uses wireless technology for wearable computing devices. WBAN consists of sensors which can be embed within human body at particular location depending upon type of sensors being used. The architecture of WBAN consists of nodes as an independent entity which is used for communication. These sensor nodes can be classified into three categories based on the roles and their usage as:

1. Implant node
2. External node
3. Body surface node

Implant nodes are placed within human body or just underneath the skin which are used for tracing information of human body. On the contrary, body surface nodes are implanted on the upper surface of body and external nodes are placed few centimeters away from patient body for extraction of information from human body. Using of IoT technology for tracing information of human body is also referred as human body area network (HBAN). These IoT technologies are used for accessing of patient data online from any place.

The management and control of network is one of the complex tasks that require robust, intelligence, and control methodologies for obtaining satisfactory performance. E-health also includes continuous monitoring and actuating mechanism. For monitoring and actuating, sensors are used for monitoring of blood pressure, heart, dopamine levels, brain activity, and actuators that pump insulin and may be used for simulating numerous body organ as pervasive elements in and on the body. These all are used for self-monitoring and facilitate healthcare delivery. Therefore, these elements around frame that act during a coordinated fashion build an individual's body

Smart Biosensors in Medical Care. http://dx.doi.org/10.1016/B978-0-12-820781-9.00007-3

space network (HBAN) blood pressure, heart, Intropin levels, brain activity, and actuators that pump hormone. This HBAN network is used for monitoring as well as for simulating various parameters for providing efficient and fast response system for healthcare management.

When dealing with large amount of data in e-health, the management and control of network becomes difficult with the use of IoT and delay sensitive applications with differing quality of service (QoS) requirements. The existing congestion control and avoidance techniques used in HBAN are not sufficient for providing efficient services in all circumstances. The congestion control mechanism used in HBAN does not include end-to-end feedback mechanism which necessitates the requirement for development of new effective congestion control algorithm. Also, there is a limitation to the control that can be accomplished from edges of the network for implementing end-to-end feedback mechanism. Thus, to enhance the performance, some additional can be implemented on routers for enhancing the end-point congestion control methods. The congestion control through routers is also termed as active queue management (AQM).

Congestion management is a mechanism to control congestion by means of determining the order of arrival of packet in interface. Management of congestion is important for formation of queues, and allocation of packets in these queues supported programming and specification of packets. Throughout lightweight traffic or transmit congestion, the traffic send to the interface is quicker [1]. Thus, congestion management facilitates in increasing queues at associate interface till interface is liberated to send them. The router maintains the priority packets to be placed in queue and the way the queues square measure maintained with regard to one another.

Sensors are devices which take environmental parameters and send it to base station for processing of that data [2]. Biosensors are devices that are used to convert biological response into electrical signals. Biosensors are finding its great use in the field of chemistry, biology, and engineering. Biosensors can be categorized into three groups based on the type of materials used by them as:

1. Biocatalyst group which comprises enzymes
2. Bioaffinity group which comprises antibodies and nucleic acid
3. Microbe group which comprises microorganism

Biosensors can be of various types as tissue based, enzyme based, thermal based, DNA biosensors, immunosensors, and piezoelectric based. These biosensors can be used in different fields.

The use of biosensors in medical field increases in recent years. Biosensors like glucose biosensors are widely used for diagnosis of diabetics. Biosensors are also used for diagnosis of infectious diseases, for this purpose a prominent technology termed as urinary tract infection (UTI) is used. Biosensors also find their usage for identification of heart failure. A biosensor based on hafnium oxide (HfO_2) can be used for detection of human interleukin (IL-10). Apart from this, there are many other biosensors which can be used for cardiac makers in undiluted serum, cancer, drug discovery, etc.

AQM is a router-based congestion control mechanism which is used for notifying the end-routers about the congestion. This congestion is detected by means of queue of the router and notifies the source before the queue overflows. There are various protocols that are based on different working of mechanism of various protocols in order to achieve stable size of queue. But all the methods and methodology are based on the concept of minimizing the size of queue within routers as small as possible [3].

This chapter proposes HBCC protocol which is a congestion control protocol for HBAN network by using AQM-based schemes which are used to provide high network utilization with low loss and delay. Thus, this chapter introduces a new fuzzy logic-based AQM control methodology which supports for efficient congestion control in HBAN network for providing smart e-health application in IoT.

2   Literature review

There are many researchers who had proposed different algorithms for congestion control for efficient working of HBAN network. Some are discussed as later.

Floyd and Jacobson [4] planned the random early detection (RED) entryways for dodging of congestion at the gateway. This detection of congestion was done by suggesting that of computing of average size of length of queue.

Jacobson [5] proposed gateway for detection of congestion by means of monitoring the length of queue and randomly dropping the packets in detection of congestion. The proposed gateway is successor to the Early Random Drop gateways, which are better than the traditional gateway in terms of measurement of average queue size, dropping packets, or marking of packets.

Yang and Reddy [6] focused on congestion control in packet switched network by using control theory. They develop a path by comparing the framework for development of new congestion control approaches.

Nejakar et al. [7] proposed RED AQM simplified model where the weighted sum of the input and output queue is used for measuring the congestion. The proposed algorithm is used to reduce average backlog of the switch in order to reduce the low speedup region.

Mahmoud et al. [8] proposed a controller technique which is used for early detection of congestion within buffer of router. This proposed technique is successor of gentle random early detection (GRED) which was based on average length of queue and delay rate for fuzzy logic-based system. The fuzzy logic system is used to present a single output which represents the probability of dropping of packet which is used to prevent and control congestion in early stage.

Braden et al. [9,10] focus on TCP congestion avoidance. They focus on controlling the congestion from edges of the network. They also propose a scheme for preventing congestion from routers.

Kamra et al. [11] propose FABA algorithm which is a rate control-based AQM algorithm which proposes QoS for providing differentiated services for efficient flow of traffic without congestion.

Ryu et al. [12] have proposed two algorithms termed as "PID-controller and Pro-Active Queue Management (PAQM)" which support end-to-end control of transmission by avoiding congestion. These protocols are designed using proportional-integral-derivative (PID) feedback control for dealing with problem to avoid congestion.

Chang and Muppala [13] proposed Q-SAPI algorithm which is PI controller-based algorithm, where adaptive PI controller is used for improving the performance of PI controller even maintaining the performance of steady state.

3   Available algorithms

AQM is defined as advancement to router-based queue management which is used for congestion detection and control by means of controlling the flow. AQM scheme drops packets or performs ECN marking for signaling the congestion at end nodes [14,15]. The congestion can be measured based on two parameters:

1. Queue based
2. Flow based

3.1   Queue based

Queue-based AQM uses size of queue as parameter for measuring the congestion. In this, the internal routers contribute for congestion control by maintaining a set of queues at each interface of routers. These interfaces hold the packets that are scheduled to start extinct from these interfaces. The congestion is maintained by dropping the packet which is set above the limit size of queue.

3.2   Flow based

This approach determines the congestion based on predicting the utilization of link and controls the congestion by taking action on the arrival rate of packets. This scheme provides early feedback for congestion.

The AQM methodology consists of various protocols based on different criteria as follows:

1. Tail drop
2. Random early detection (RED)
3. Random exponential marking (REM)
4. Blue queue
5. REM controller
6. Adaptive virtual queue (AVQ)

Tail drop: Tail drop uses default congestion control techniques of avoiding congestion by dropping packets arriving after the queue reaches to its maximum threshold limit or after the queue is filled. Thus, it drops the packets arrived after the output queue is filled with packets. Drop tail suffers with three important drawbacks which are as follows:

Lockout: Lockout is a synchronization process for congestion control which is being used by drop tail for avoiding congestion. It allows single connection or few flows to control queue space and thus prevents other connections from getting room in queue.

Full queue: Tail drop is used to prevent congestion as queue becomes full, so managing the size of queue can be used for management of congestion control.

Global TCP synchronization: Global synchronization may be a cycle of underutilization following the drop burst followed by amount of overload. It happens as a result of the burst of drops ends up in an oversized variety of TCP sources reducing their window size at constant time.

The tail-drop algorithm is not an efficient AQM algorithm and suffers with the problem of link efficiency, low throughput, and bad fairness.

Random early detection (RED): RED algorithm was proposed by Floyd and Jacobson [4] and Floyd et al. [16], which is an efficient algorithm for avoidance of congestion in routers and gateways. RED algorithm is recommended by IETF in RFC 2309. It uses length of queue as measure for congestion control. The notification of congestion can be determined by dropping of packets or making of ECN in case packets are capable of ECN. The computation of occupancy of time-average queue is done through exponential averaging scheme which is shown later:

$$\text{Avg}(t+1) = (1 - \text{wg}).\text{Avg}(t) + \text{wqB}(t)$$

Here, wq = average time constant

$B(t)$ = instantaneous queue occupancy

On exceeding the minimum threshold of average queue length (AQL), this generates a notification of congestion which is proportional to excess queue length.

RED algorithm uses low pass filter for calculating the average size of queue. For efficient management of congestion, this size of queue is compared with maximum and minimum threshold of queue. It works on the concept that if size of queue is less than minimum threshold (queue size < minimum threshold) then in such case no packet is marked. Alternatively, if the size of queue is more than maximum threshold then it marked all the packets trending toward gateway. In order to maintain the size of queue within the limit, these marked packets are dropped and thus maintain the congestion within the network [17].

Random exponential marking (REM): This algorithm is used to decouple congestion from performance measure. Congestion measure represents excessive use of bandwidth, whereas performance measure is used to indicate the length of queue and delay within the network. REM is advanced version of RED, which uses AQL as parameter for detection of congestion within the network, whereas REM uses detection of mark policy for detection of congestion within the network which is better in terms of performance than RED algorithm [18].

REM algorithm maintains the length of queue around a small target by measuring the capacity of link with respect to the rate of input from all the users and this is achieved irrespective of numbers of user who are sharing the link. This algorithm maintains rate mismatch and queue mismatch, where the rate mismatch is the difference between input rate and capacity of the link whereas the queue mismatch is the difference between current length of queue and targeted length of queue which is updated at regular intervals. REM increases probability of marking with increase in number of users, which will also increase the congestion within network and thus can control the occupancy of queue around the target queue length. REM uses sum of prices of the link along a path for measuring the congestion along a path.

Blue queue: It uses hybrid control scheme to overcome with the problems of RED algorithm by making use of queue size congestion measuring scheme. This algorithm makes use of flow and size of queue as hybrid parameter for modifying the rate of notification of congestion. It uses queue congestion and utilization of link for determining the rate of loss of packet. Thus, blue algorithm focuses on packet loss rather than length of queue as used by RED algorithm.

BLUE algorithm uses probability parameter for marking and dropping of packets. If queue drops packets due to overflow of buffer, then the BLUE algorithm increases the probability parameter, which increases the rate of sending notification for congestion or dropping of packets. Alternatively, if link is idle or queue is empty, it lowers the rate of sending notification of congestion and dropping of packets by decreasing the parameter of probability. Thus, using this algorithm we can get the correct rate of sending up of notification of congestion as well as dropping of packets [19].

Adaptive virtual queue (AVQ): AVQ uses implicit marking probability and thus there is no requirement of random number generation. This algorithm also replaces calculation of marking probability with computation of capacity of virtual queue and thus better than other algorithms in terms of loss, delay, and utilization.

It uses virtual queue with capacity less than the actual capacity of the link and this virtual queue is updated every time the packet arrives in the queue in order to reflect the new arrival. It drops or marks the packet from the real queue as the virtual buffer gets overflows. The virtual capacity of each link is adapter in nature in order to ensure that each link achieves maximum utilization of link with the total input flow of packets.

Thus, it can be concluded that AVQ algorithm is rate-based marking approach which performs better than previously introduced algorithm which uses length of queue and average length of queue for marking. AVQ algorithm performs better than traditional algorithms for varying condition of changing network in terms of parameters like fairness, utilization, and AQL.

Table 7.1 gives the comparative analysis of AQM-based algorithms proposed by different researchers on different parameters.

Table 7.1 shows the comparison analysis of various AQM protocol in terms of their parameters and their working [23].

Table 7.1 Comparative analysis of existing algorithms.

Year	Algorithms	Approach	Congestion Control Measures	Performance Metrics	Full Form
1993	RED	Avoiding congestion at gateway	Average length of queue	Global synchronization, fairness, loss of packet, and delay of queue	Random early detection
1999	REM	Optimize internet congestion control [20]	Average length of queue, input rate	Loss of packet, buffer occupancy, fairness, and goodput	Random early marking
2000	CHOKe	Queue management for fair bandwidth allocation	Average length of queue	Throughput, minimum overhead, easy to implement	CHOse and keep for responsive flows
2001	DRED	Queue management	Instantaneous length of queue	Utilization of link and rate of packet loss	
2001	ARED	Queue management	Length of queue	Utilization and delay	Adaptive random early detection
2001	AVQ	Active queue management	Rate of input	Loss of packet, utilization of link, length of queue, response to changing network condition	Adaptive virtual queue
2001	SFB	Queue management for fairness	Loss of packets, utilization of link	Throughput, length of queue, delay	Stochastic fair blue
2002	PI	Router management for TCP flow	Length of queue	Queue utilization, robustness, queuing delay	Proportional integral
2003	CARE	Queue management		Throughput, fairness, stability, adaptability, complexity	Capture recapture model
2004	FABA	Queue management for rate control	Rate of input	Complexity, fairness, scalability, throughput	Fair adaptive bandwidth allocation
2005	Yellow	Queue Management	Rate of input	Loss of packet, robust, utilization of link, stability, queue delay	
2005	E-RED	AQM protocol for low- and high-speed TCP protocols	Length of queue	Throughput, stability, and queuing delay	Exponential-RED
2005	RAQM	Rate-based algorithm	Rate of input	Fast response, goodput, stability	

(Continued)

Table 7.1 Comparative analysis of existing algorithms. (*Cont.*)

Year	Algorithms	Approach	Congestion Control Measures	Performance Metrics	Full Form
2005	DC-AQM	Queue management	Length of queue	Utilization of link and length of queue	
2006	Q-SAPI	Queue-based adaptive controller	Length of queue	Loss of packet, queuing delay, and utilization of link	Q-stable adaptive proportional-integral
2007	RaQ	Queue management	Rate of input and length of queue	Queue length, stability, convergence, robustness	Robust active queue management scheme
2007	LRED	Measurement of packet loss ratio	Instantaneous length of queue and packet loss ratio	Response time, robustness, flexible system configuration	
2007	PAQM	Queue management	Length of queue and rate of input	Low queuing delay, utilization of link, complexity, and easy configuration	Proactive queue management
2008	IMC-PID	AQM controller based for large delay network	Length of queue	Stability, robustness, convergence, utilization of link, delay, and jitter	
2008	AOPC	Controller based on optimization	Length of queue and packet loss ratio	Responsiveness, utilization, delay, convergence rate, robustness	Adaptive optimized proportional controller
2010	RRED	Queuing discipline for network scheduler, meant for long-term traffic pattern	Security based	TCP throughput for LDoS attack [21,22]	Robust RED algorithm
2011	HSTCP-H2	Controller based on large bandwidth-delay product networks	Length of queue and rate of input	Goodput and intra-protocol fairness	High-speed TCP modeling
2013	RC			Robustness, queue stability	

4   Fuzzy logic

Fuzzy logic is also known as computational intelligence which is used to control methods for communication data networks. Fuzzy logic is used in classical control system which is used to alleviate the complex parameters of system into simpler mathematical models that can be used for simplifying the model or designing of model traceable for controller design. Fuzzy logic has been used in recent years for preventing and solving the problem of congestion control. Fuzzy logic control (FLC) is used at many situations where obtaining of analytical model is not easy or the model is very complex and highly nonlinear in nature. FLC can be used for capturing the attributes of control system based on observable phenomena.

Our proposed algorithm or proposed design consists of hybrid of all the algorithms discussed earlier (RED, REM, BLUE, REM controller, PI controller, and AVQ) which form a new model which is integrated with concept of fuzzy logic to develop a new algorithm which will be beneficial than the existing algorithm.

4.1   Fuzzy logic-based congestion control (FBCC) protocol

Proposed algorithm is a combination of all the earlier approaches with concept of fuzzy logic embed with them for improved performance of the algorithm for detection of congestion within network (Fig. 7.1). The proposed algorithm uses fuzzy interface process (FIP) which is used for detection of congestion, whereas FLC controller is used as expert system which is knowledge-based decision system for efficient detection of congestion. FLC is used to process input and

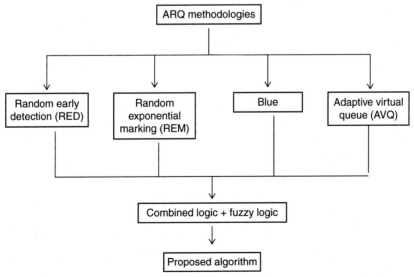

FIGURE 7.1 Terminology for FBCC protocol.

produce output using fuzzy linguistic rules. The process of FLC consists of following stages as:

1. Fuzzification
2. Evaluation of rules
3. Aggregation of outputted rules
4. Defuzzification

The algorithm is used to control congestion as congestion control calculates dropping probability for calculating AQL and queuing delay (D). There are four steps which are used for implementation of algorithms:

- Step 1: fuzzification
- Step 2: evaluation of rules
- Step 3: aggregation of output rules to single fuzzy set
- Step 4: defuzzification of output linguistic variable to crisp values

Fuzzification: Fuzzification is process of input crisp values to input linguistic values using following steps:

1. Fuzzy set range for each input linguistic value based on universe of discourse
2. Crisp value represents numerical value

Evaluation of rules: It changes the fuzzified input variables into membership degree for every linguistic rules based on fuzzy set rules using following steps:

1. Fuzzification of input variables is evaluated and applied on antecedent part of rules
2. This antecedent part is processed
3. Every consecutive part of rule is evaluated by obtaining membership degree of output variable
4. Multiple antecedent rules are found and computation of these rules is calculated using fuzzy set operations
5. Result of antecedent rules helps for calculation of membership degree for every output from linguistic rules

Aggregation of output rules into single fuzzy set: This step is used to aggregate different sets into a single fuzzy set based on concept of fuzzy logic using following steps:

1. Membership degree obtained from step 2 for every output is aggregated into a single output rule. This single output is termed as single fuzzy set
2. This step is useful for calculating fuzzy set for every output variable

Defuzzification of output linguistic variable to crisp values: This step converts the fuzzy set into output linguistic variable using following step:

1. Based on fuzzy set FIP generates crisp value for each output linguistic variable

Fuzzy set: In the proposed approach each fuzzy set is associated with each linguistic variable in FLC. These fuzzy sets are chosen based on the behavior of input linguistic variables. Thus, if input linguistic variable is low then the mean average queuing delay will also be low. These fuzzy sets are associated with parameters of AQL and delay. As the fuzzy set is identified from input linguistic variables, the corresponding membership function is generated. These membership functions are used to form different shapes based on the type of problems. These shapes are used to figure out the continuous domain of each control variable which is used to trace a predictable system. This can also be used to find linguistic values for specific input variable and can also be used to trace exact values for controlled system's input. Thus, it proposes a fuzzy-based decision-making process which can take action based on input and output parameters.

In order to define the rule-base in fuzzy logic, following points should be kept in mind:

1. All conditions of a system should be considered and taken into consideration
2. Rule-base should not contain any illogicality, that is, it should be consistent

Thus, fuzzy controller works on the control rules in following manner:
Consideration (linguistic variables as AQL (A_1) and packet delay (D))
{
If (A_1 == conservative && D == average)
Then probability of packet drop (Pd) = 0;
Else if (A_1 == aggressive and D == long)
Then output == High for quick response
}
Some other conditions can be as follows.
Table 7.2 shows the condition and action taken against different inputs from linguistic variables for avoiding congestion in router.

Table 7.2 Conditions of fuzzy system with different inputs.

A_1 (Attribute)	D (Attribute)	Pd (Attribute)
Conservative	Little	0
Conservative	Average	0
Conservative	Long	0
Middle	Little	0
Middle	Average	0
Middle	Long	Low
Aggressive	Little	0
Aggressive	Average	Moderate
Aggressive	Long	High

Thus, the linguistic variables like A_1 and D are used by controller for finding the desired output. Thus, these fuzzy rules are used for obtaining the control signals based on congestion in buffer of router. This logic is implemented by means of expert system for managing the congestion within router.

Once these all are done, the membership degree of THEN part of rules is aggregated into single fuzzy set and then finally the defuzzification is done where single aggregate fuzzy set of output variable is used for calculating crisp value for probability of packet drop from the routers.

5 Application of FBCC in HBAN system for smart e-health application using biosensors

Medical science is advancing day by day; now the use of internet-based technology and the use of sensors for measurement of various parameters of the patient is a general term. Use of biosensors for monitoring various parameters in a human body is performed in various papers by various researchers. In order to perform the smart e-health, the transfer of these sensor data to the main server should be done in a fast and efficient manner. There are various protocols proposed in HBAN system for communication purpose to remove congestion so that the data should pass in an efficient manner. In this chapter, we had proposed for a new approach for data transfer from the biosensors to the main server, where the data are processed. The proposed algorithm FBCC uses fuzzy-based approach which is helpful in setting congestion-free communication from biosensors to main server (Fig. 7.2).

In Fig. 7.2, the biosensors manage to capture various body parameters like temperature, heartbeats, nerves, etc., and this information is sent to router where FBCC algorithm runs using concept of fuzzy logic discussed earlier. The FBCC algorithm will help in preventing the congestion for the data receiving from biosensors. The router will fetch the information and send it to server where the information is stored and processed to be use by the doctors and for intelligent systems.

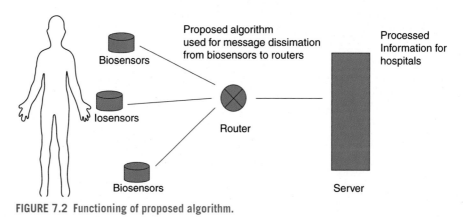

FIGURE 7.2 Functioning of proposed algorithm.

6 Conclusion

This chapter discusses about congestion control in HBAN for smart e-health application by making use of biosensors. For achieving the high throughput for smart e-health, the communication from biosensors to routers is one of the prime aspects and this chapter focuses on congestion-free communication from biosensors to routers. This chapter proposes an algorithm-using concept of fuzzy logic to enhance the performance of e-health application for HBAN network. This algorithm is based on the concept of taking parameters of length of queue and delay of packets as parameters for proving the work of the proposed protocol. The algorithm is proposed by taking all the conditions necessary for the working of HBAN network and based on all the condition different approaches are taken into consideration which help in proving the durability and reliability of the proposed algorithm. The approach proposed will also paved way for future work in the direction of HBAN network using biosensors which will also help in reducing energy and lead a pathway for future network.

References

[1] A. Dumka, Smart information technology for universal healthcare, in: Healthcare Data Analytics and Management, Elsevier, 2018, pp. 211–226.

[2] A. Dumka, S. Chaurasia, A. Biswas, H.L. Mandoria, A Complete Guide to Wireless Sensor Networks from Inception to Current Trends, first ed., CRC Press, 2019.

[3] L. Long, J. Aikat, K. Jeffay, F. Smith, The effects of active queue management and explicit congestion notification on web performance, IEEE/ACM Transact. Netw. 15 (2005) 1217–1230.

[4] S. Floyd, V. Jacobson, Random early detection gateways for congestion avoidance, IEEE/ACM Transact. Netw. 1 (4) (1993) 397–413.

[5] V. Jacobson, Congestion avoidance and control, Comput. Commun. Rev. 18 (4) (1998) 314–329.

[6] C. Yang, A. Reddy, A taxonomy for congestion control algorithms in packet switching networks, IEEE Netw. Mag. 9 (4) (1995) 34–45.

[7] S.M. Nejakar et al., Development of modified RED AQM algorithm in computer network for congestion control, Development 1 (8) (2014) 380–385.

[8] Mahmoud, et al. Fuzzy logic controller of gentle random early detection based on average queue length and delay rate, Int. J. Fuzzy Syst. 16 (1) (2014) 9–19.

[9] B. Braden et al., Recommendations on queue management and congestion avoidance in the internet, Technical Report, RFC 2309, USA, 1998.

[10] B. Braden et al., RFC 2309—recommendations on queue management and congestion avoidance in the internet, Technical Report, IETF RFC, USA, 1998.

[11] A. Kamra, S. Kapil, V. Khurana, V. Yadav, H. Saran, S. Juneja, R. Shorey, SFED: a rate control based active queue management discipline, IBM India Research Laboratory Research report # 00A018, 2000.

[12] S. Ryu, C. Rump, C. Qiao, Advances in active queue management (AQM) based TCP congestion control, Telecommun. Syst. Model. Anal. Design Manag. 25 (3–4) (2004) 317–351.

[13] X. Chang, J.K. Muppala, A stable queue-based adaptive controller for improving AQM performance, Comput. Netw. 50 (13) (2006) 2204–2224.

[14] C. Zhu, O.W.W. Yang, J. Aweya, M. Ouellette, D.Y. Montuno, A comparison of active queue management algorithms using OPNET Modeler, OPNET 2002 30 (6) (2002) 158–167.

[15] J. Aweya, M. Ouellette, D.Y. Montuno, A control theoretic approach to active queue management, Comput. Netw. 36 (2–3) (2001) 203–235.

[16] S. Floyd, R. Gummadi, S. Shenker, Adaptive RED: an algorithm for increasing the robustness of RED's active queue management, AT&T Center for Internet Research at ICSI, 2001 (under submission).

[17] C.V. Hollot, V. Misra, D. Towsley, W. Gong, A control theoretic analysis of RED, in: Proceeding of INFOCOM 2001, IEEE, Anchorage, AK, USA, 2001, pp. 1510–1519.

[18] S. Athuraliya, V.H. Li, S.H. Low, Q. Yin, REM: active queue management', IEEE Netw. Mag. 15 (2001) 48–53.

[19] W. Feng, K.G. Shin, D.D. Kandlur, D. Saha, The BLUE active queue management algorithms, IEEE/ACM Transact. Netw. (10) (4) (2002) 513–528.

[20] S. Athuraliya, D. Laspsley, S.H. Low, An enhanced random early marking algorithm for internet flow control, in: Proceedings of Infocom, IEEE, Israel, 2000.

[21] Z. Changwang, Y. Jianping, C. Zhiping, C. Weifeng, RRED: robust RED algorithm to counter low-rate denial-of-service attacks, IEEE Commun. Lett. 14 (5) (2010) 489–491.

[22] H.V. Shashidhara, S. Balaji, Low rate denial of service (LDoS) attack—a survey, Int. J. Emerg. Technol. Adv. Eng. 4 (6) (2014).

[23] T. Bhaskar Reddy, A. Ahammed, R. Banu, Performance comparison of active queue management techniques, Int. J. Comput. Sci. Netw. Secur. 9 (2) (2009) 405–408.

Biofabrication of graphene QDs as a fluorescent nanosensor for detection of toxic and heavy metals in biological and environmental samples

Fahmida Khan, Subrat Kumar Pattanayak, Padma Rani Verma, Pradeep Kumar Dewangan

Department of Chemistry, National Institute of Technology, Raipur, India

1   Introduction

Heavy metal ions are ubiquitous distributed nonbiodegradable substances [1,2] that lead to a greater risk to human health by its accumulation in the human body through air [3], beverages [4], vehicle emissions [5], batteries [6], food chain [7], and industrial activities [8] in which water plays a key role. Exposure of heavy metal ions such as lead (Pb), mercury (Hg), uranium (U), cadmium (Cd), etc., even at low concentration and some essential nutrient metal ions such as iron (Fe), zinc (Zn), cobalt (Co), etc., at high concentration are toxic for the human health as it causes cardiovascular diseases, cancer mortality, neurological disorders [9,10], and are toxic to the environment as well. Therefore, detection of these metal ions at low concentration is prior for the environmental protection and for the prevention of diseases.

The great efforts have been developed for sensitive and selective detection techniques of toxic heavy metal ions. Atomic absorption spectrometry (AAS) techniques [11] and inductively coupled plasma mass spectrometry (ICP-MS) techniques [12] were reported earlier for heavy metal analysis. The different electrochemical techniques such as voltammetric and potentiometric techniques were also reported for heavy metal detection [13,14]. However, stability, selectivity, detection limits, compatibility with aqueous environment, and ease of sampling remain as significant challenges for these techniques, which are properly resolved by advanced technologies of electronics [15], electrochemistry [16], and optics [17]. Of them, optical detection techniques, that is, fluorescent sensor, have gained importance in recent

Smart Biosensors in Medical Care. http://dx.doi.org/10.1016/B978-0-12-820781-9.00008-5

years for detection of heavy metal ions with simplicity, high specificity, and low-detection limits. Several optical sensors by using different materials such as fluorescent aptamers [18], porphyrins [19], DNAzymes [20], organic dyes [21], metal-organic frameworks [22], and quantum dots [23] were reported as fluorescent sensor for the detection of heavy metal ions.

In recent years, quantum dots (QDs) are the emerging field of nanotechnology that has gained major importance because of its variable applications with specificity. QDs are the semiconductor nanoparticles whose all three dimensions found to be below 10 nm. Various types of QDs have been reported from the various groups of the periodic table such as from Group II—Zn, Cd; Group VI—S, Se, etc. QDs reported from the carbon family as graphene quantum dots (GQDs). Graphene is a two-dimensional carbon nanomaterial with nonluminescent property. However, when graphene transforms into zero-dimensional graphene quantum dots, they exhibit the luminescence property with various optical and electronic characteristics (Fig. 8.1).

GQDs and its functionalized derivative have gained a special attention because of its applicability in bioimaging [24], light emitting diodes [25], photovoltaic devices [26], and heavy metal ions detection [27]. GQDs exhibit the various applications due to their intrinsic superiorities such as excellent biocompatibility, high photoluminescence (PL), low toxicity, high solubility in various solvents, and good resistance to photobleaching [28–31]. GQDs show enhanced optical and electronic characteristics due to the effect of quantum confinement, surface functionalization, edge effect, and heteroatom doping [32–35]. Doping the graphene with heteroatoms such as H, N, S,

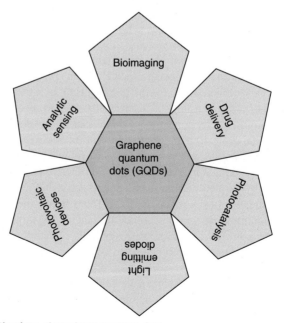

FIGURE 8.1 Applications of graphene quantum dots.

F, Mg, and Cl alters its intrinsic performance effectively through a tunable bandgap in the energy spectrum of graphene [36–40]. Several methods have been proposed for the synthesis of nitrogen-doped GQDs (N-GQDs) [41–43] but their fluorescence (FL) quantum yields (QYs) are usually less than 25%. Therefore, it is desirable to synthesize such N-GQDs with high FL QYs and with excellent PL characteristics.

In this chapter, we reported the one-step facile microwave synthesis of blue bright fluorescent N-GQDs by using glucose and urea as source materials for the detection of heavy metal ions. This method does not require any expensive reagents, thus making it environment-friendly and cost-effective approach. This synthesized N-GQDs demonstrates the excellent solubility in water, intense blue emission, and PL characteristics with uniform size of GQDs.

2   Experiment

Instrument and reagents: The chemicals used were glucose (99%), urea (99%) phosphoric acid (85wt% in water), and ethanol purchased from Sigma-Aldrich. These chemicals were used without any further purification and double distilled deionized water was used. UV-Visible Spectrophotometer (Labtronics) Model LT-2900 was used to characterize the sample for UV spectral analysis. Fluorescence intensity and the maximum excitation and emission wavelengths of N-GQDs were measured by Cary Eclipse Fluorescence Spectrophotometer of Agilent Technologies. High-resolution transmission electron microscopy (HR-TEM) micrograph was taken from JEOL 2012 instrument. Microwave radiation synthesized GQDs were carried out in CEM Discover Microwave Synthesizer instrument.

Preparation and optimization of the graphene quantum dots: Added 2 g of glucose with 0.5 g of urea in the presence of 10 mL of phosphoric acid in a 25 mL round bottom flask. Now, the reaction was refluxed in microwave synthesizer for 20 min at 170°C temperature. The dark brown color residue obtained was diluted with 50 mL of double distilled water and centrifuged at 13,000 rpm for 10 min. Now, the yellowish-brown color of solution was obtained with blue photoluminescent features (Fig. 8.2).

3   Results and discussion

The UV-Vis absorption spectrum of the synthesized N-GQDs shows the absorption peak at 277 nm and 379 nm, respectively. The sharp absorption peak at 277 nm corresponds to π-π* transition for the aromatic sp^2 domain, while the other broad absorption peak at 379 nm corresponds to n-π* transition for the presence of double bonds of carbon and heteroatoms at the edges of the graphitic sheets (Fig. 8.3).

The formation of N-GQDs was also characterized using Fourier-transform infrared (FTIR) spectra shown in Fig. 8.4. The broadband at 3449 cm^{-1} ascribed for the O-H and N-H vibration modes which are involved in hydrogen bonds. The band at

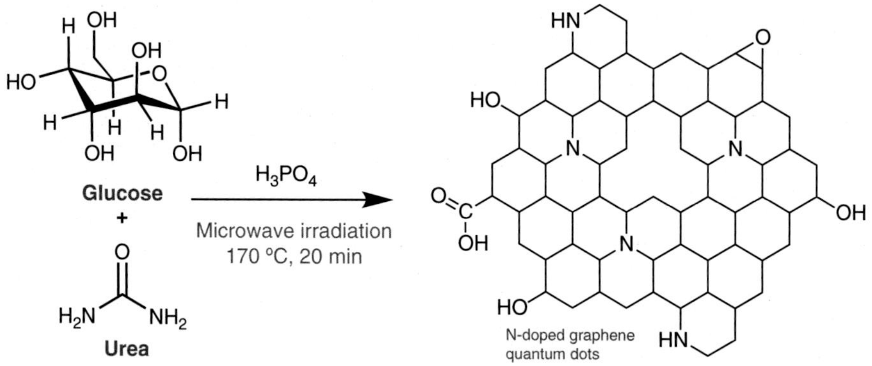

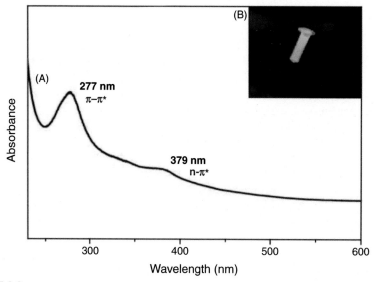

FIGURE 8.2 Schematic representation of synthesis of nitrogen-doped graphene quantum dots (N-GQDs).

FIGURE 8.3

(A) UV-Vis spectrum of synthesized N-doped graphene quantum dots and inset, and (B) the emission of *blue* bright luminescence by 365 nm UV lamp.

2367 cm^{-1} is the characteristic for C-H vibration and the band at 1635 cm^{-1} for the $-C = O$, $-C = N$ stretching vibration modes. The band at 1184 cm^{-1} and 1005 cm^{-1} is responsible for C-C, C-O, C-N bond. The weak band at 530-409 cm^{-1} contributes for bending vibrations.

The Raman spectrum depicted in Fig. 8.5 shows the two distinctive peaks for the graphitic structures. The D-band obtained at 1258 cm^{-1} due to the longitudinal vibration of disorder carbon, and G-band obtained at around 1896 cm^{-1} referred to the transverse graphene vibrational mode (out-phase). The atomic ratio of sp^2/sp^3

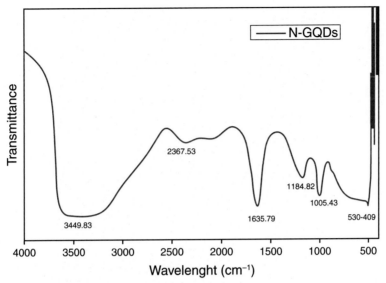

FIGURE 8.4 FTIR spectral analysis of synthesized N-GQDs.

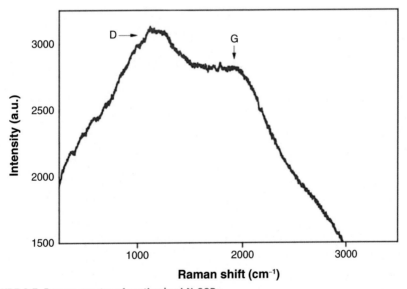

FIGURE 8.5 Raman spectra of synthesized N-GQDs.

carbons, expressed as the ratio of the intensity of Raman peaks (ID/IG), resulted to be 1.09. The blue shifts and the broadening of both D- and G-bands obtained possibly due to the doping of heteroatoms like nitrogen [44,45].

HR-TEM of N-GQDs confirms the spherical morphology of the individual GQDs size between 3 nm and 7 nm and N-GQDs particles form monodispersed layer as shown in Fig. 8.6.

FIGURE 8.6 High-resolution transmission electron micrograph (HR-TEM) of nitrogen-doped graphene quantum dots.

The fluorescence (FL) emission spectrum of the N-GQDs solution is broad. To explore the PL characteristics of the obtained N-GQDs, a detailed investigation of the FL by changing the excitation wavelengths ranging from 200 nm to 300 nm was carried out in Fig. 8.7. The emission peaks lie in the range of wavelength 355–435 nm with increasing excitation wavelength. This excitation PL behavior is different from most FL carbon-based materials [46–48], which implies that both the size and the emission sites of the sp^2 clusters contained in N-GQDs should be uniform and shows in the inset photos (Fig. 8.3B), and the N-GQDs exhibit intense blue emission under a 365 nm UV lamp chamber, which is different from the colorless emission under visible light (400–780 nm). Thus, its blue PL effect is proposed to be produced by the ordered sp^2 clusters because the N-GQDs contain abundant sp^2 structures and oxygen-containing functional groups (such as C–O, C = O, and –COOH) [49].

3.1 The mechanism of binding of cadmium ion by N-GQDs

Fluorescence properties of N-GQDs are used as chemosensor for cadmium ion. Quenching property of intensity of fluorescent N-GQDs in the presence of metal ion is basic principle for detection of metal ions. It is observed that the fluorescence emission intensity N-GQDs were decreased regularly by addition of Cd(II) ions. This phenomenon indicates the binding occurs between Cd(II) and N-GQDs, which is responsible for quenching of fluorescence intensity. N-GQDs are highly functionalized by hydroxyl group, amine group, carboxylic acid, and nitrogen atom that replaced some carbon atom, which is to promote formation of coordination bond between Cd(II) and N-GQDs. The fluorescence intensity of Fig. 8.8 shows that fluorescence

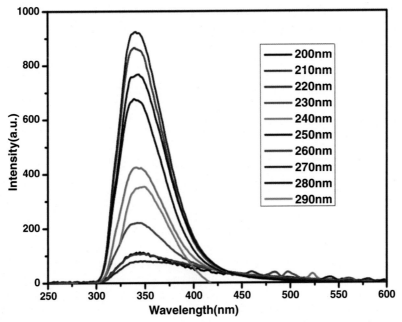

FIGURE 8.7 Fluorescence (FL) emission spectra of the N-GQDs for different excitation wavelength from 200 nm to 300 nm.

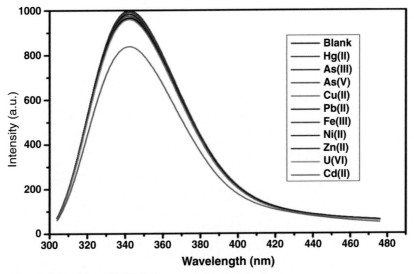

FIGURE 8.8 Selectivity of Cd(II) in presence of various ion.

intensity of N-GQDs is effectively quenched by Cd(II) ion. The fluorophore may be attributed in the excited state is non-radiative transfer of electron to the d-orbital of Cd(II) ions.

The linear graph shows good correlation ($R^2 = 0.987$) under the concentration range from 0 to 80 ppb.

The limit of detection on the basis of 3 times the standard deviation rule (LOD = 3Sd/s) is 0.64 ppb. The obtained LOD of N-GQDs sensor for Cd(II) shows much lower than previously reported methods.

3.2 The selectivity of the chemosensor for Cd(II) detection

To evaluate the selective detection of Cd(II) ion in the present sensing method ion water sample, 50 ppb various metal ions [Hg(II), As(III), As(V), Cu(II), Pb(II), Fe(III), Ni(II), Zn(II), U(VI), Cd(II)] were added into N-GQDs at 8.0 pH and reaction time, 4 min for this purpose, and the fluorescence spectra were then recorded. As observed in Figs. 8.8 and 8.9, significant fluorescence quenching was shown in the presence of Cd(II) ion while in the presence of other ions quenching effect is almost negligible. Fig. 8.8 clearly shows that Cd(II) ion significantly quenched in fluorescence intensity of N-GQDs than other ion, so we can selectively used N-GQDs for the detection of Cd(II) in environmental water. To examine the selectivity, further in another experiment 50 ppb of Cd(II) separately and 50 ppb of Cd(II) mixed with 50 ppb of above mention metal ions were also added into N-doped GQDs aqueous solution, respectively. Then fluorescence spectra recorded and examine quenching effect of Cd(II) and mixture of Cd(II) and M^{n+}. Result shows that quenching effect of N-GQDs in the presence of Cd(II) ion does not interfere mixture of metal–metal ions. It is indicated that N-GQDs are highly selective for detection of Cd(II) ion.

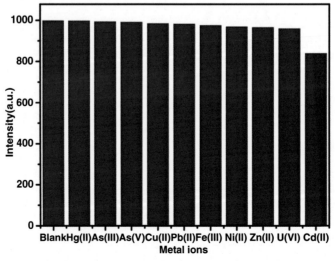

FIGURE 8.9 Column representation of selectivity of metal ions.

3.3　Optimization of GQDs as a fluorescence sensor for detection of ions

For controlled experiment of detection of Cd(II) in environmental water sample by N-GQDs that were performed under best experimental condition, pH, reaction time, and concentration of the Cd(II) ion are important factors and are shown in Figs 8.10–8.13. These parameters are affect detection of metal–metal ion and QY.

The pH is an important parameter for the selectivity of metal ions by N-GQDs, which affect the fluorescence intensity. Therefore, here it shows that effects of pH on the quenching intensity of N-GQDs in the solution of Cd(II) ion were studied by changing of pH of solution from 4.0 to 11 pH, and the fluorescence intensity decreases regularly from 4.0 to 8.0 pH. After 8 pH fluorescence intensity increases regularly. So at 8.0 pH maximum quenching occurs. These data indicate that slightly basic medium favors binding of Cd(II) and N-GQDs. Therefore, pH 8.0 is favorable pH for further experiment.

In another controlled experiment, reaction time for examine of fluorescence spectra of N-GQDs is important. For binding of metal ion with GQDs some time is required; therefore, fluorescence spectra are taken at various time interval for best result. The surface of GQDs interacts Cd(II) during the reaction time. In this method, intensity of GQDs is examined in the presence of Cd(II) ion at pH 8.0 at a different time interval (0–8 min). The result indicates that quenching of fluorescence intensity

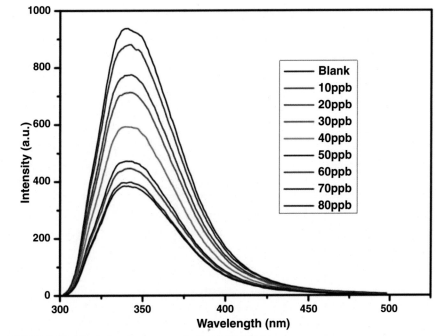

FIGURE 8.10 Effect of various concentration of Cd(II) ion.

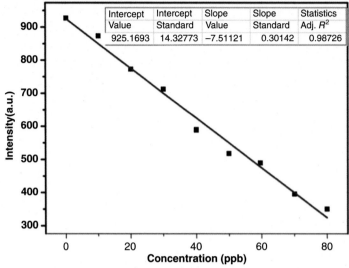

Intercept Value	Intercept Standard	Slope Value	Slope Standard	Statistics Adj. R^2
925.1693	14.32773	−7.51121	0.30142	0.98726

FIGURE 8.11 Linear graph of quenching intensity N-GQDs as a function of concentration of Cd(II) ion.

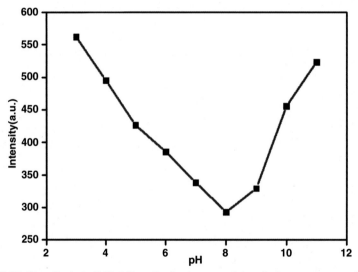

FIGURE 8.12 The effect of pH (3–11) on the fluorescence intensity in an aqueous solution of GQDs with Cd(II) ion.

of aqueous solution GQDs deceases regularly from 0 to 4 min. After that fluorescence intensity remains constant. Therefore, for further experiment 4 min is chosen.

Effect of concentration of Cd(II) ion is observed by various concentration of Cd(II) that is added into GQDs. In this purpose, 40 μL of 0.01 M of N-GQDs at pH 8.0 solution is taken in voil with different concentration. In this method, 0–80 ppb of

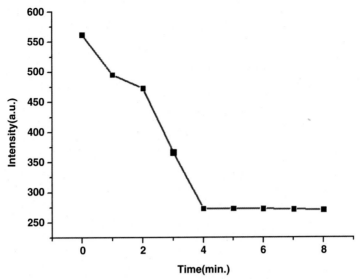

FIGURE 8.13 The response time curve to Cd(II) ion at 270 nm in ultrapure water at pH 8.

metal is taken with mixture of GQDs. After 4 min reaction time fluorescence spectra are taken which shows that quenching in intensity increases regularly with concentration of Cd(II) increases in mixture of solution. This result shows that Cd(II) acts as quencher for N-GQDs. So this method is highly selective for determination and detection of cadmium ions.

4   Conclusion

A facile microwave-assisted one-pot synthesis of N-GQDs from glucose was developed. The PL, micromorphology, chemical structure characteristics, and selectivity of synthesized N-GQDs were reported. The size of the N-GQDs is found to be around 3–7 nm. The N-GQDs exhibit strong blue light emission under 365 nm UV light. This characteristic behavior of microwave synthesized N-GQDs will open up avenues in the fields of bioimaging, biosensors, drug delivery, cancer treatment, and environmental remediation. In this chapter, we have discussed how the detection of heavy metal ions is carried out by the GQDs nanosensor. Our synthesized N-GQDs show best selective for toxic heavy metal Cd(II) ion detection, and can be best applicable of other sectors also.

References

[1] L. Maria, Biosensors for Environmental Applications, 2011, pp. 1–16. Available from: http://www.intechopen.com/books/environmental-biosensors/biosensor-for-environmental-applications.

[2] A. Singh, R.K. Sharma, M. Agrawal, F.M. Marshall, Health risk assessment of heavy metals via dietary intake of food stuffs from the wastewater irrigated site of a dry tropical area of India, Food Chem. Toxicol. 48 (2010) 611–619, doi: 10.1016/j.fct.2009.11.041.

[3] A.T. Jan, M. Azam, K. Siddiqui, A. Ali, I. Choi, Q.M.R. Haq, Heavy metals and human health: mechanistic insight into toxicity and counter defense system of antioxidants, Int. J. Mol. Sci. 16 (2015) 29592–29630.

[4] P. Senthil Kumar, A. Saravanan, Sustainable wastewater treatments in textile sector, in: Sustainable Fibres and Textiles, Woodhead Publishing, 2017, pp. 323–346.

[5] G.C. Lough, J.J. Schauer, J.-S. Park, M.M. Shafer, J.T. Deminter, J.P. Weinstein, Emissions of metals associated with motor vehicle roadways, Environ. Sci. Technol. 39 (2005) 826–836.

[6] S. Yuliusman, A. Nurqomariah, R. Fajaryanto, Recovery of cobalt and nickel from spent lithium ion batteries with citric acid using leaching process: kinetics study, in: Proceedings of the E3S Web of Conferences, Vol. 67, Berdyansk, Ukraine, 2018.

[7] E.M. Suszcynsky, J.R. Shann, Phytotoxicity and accumulation of mercury in tobacco subjected to different exposure routes, Environ. Toxicol. Chem. 14 (1995) 61–67.

[8] A. Arruti, I. Fernández-Olmo, A. Irabien, Evaluation of the contribution of local sources to trace metals levels in urban PM2.5 and PM10 in the Cantabria region (Northern Spain), J. Environ. Monit. 12 (2010) 1451–1458.

[9] H. Hussein, et al. Tolerance and uptake of heavy metals by pseudomonads, Process Biochem. 40 (2) (2005) 955–961.

[10] X. Yang, Y. Feng, Z. He, P.J. Stoffella, Molecular mechanisms of heavy metal hyperaccumulation and phytoremediation, J. Trace Elem. Med. Biol. 18 (2005) 339–353.

[11] R. Kunkel, S.E. Manahan, Atomic absorption analysis of strong heavy metal chelating agents in water and waste water, Anal. Chem. 45 (8) (1973) 1465–1468.

[12] S. Caroli, et al. Determination of essential and potentially toxic trace elements in honey by inductively coupled plasma-based techniques, Talanta 50 (2) (1999) 327–336.

[13] E. Bakker, Y. Qin, Electrochemical sensors, Anal. Chem. 78 (12) (2006) 3965–3984.

[14] B.J. Privett, J.H. Shin, M.H. Schoenfisch, Electrochemical sensors, Anal. Chem. 80 (12) (2008) 4499–4517.

[15] A.M. Simões Da Costa, I. Delgadillo, A. Rudnitskaya, Detection of copper, lead, cadmium and iron in wine using electronic tongue sensor system, Talanta 129 (2014) 63–71.

[16] T. Alizadeh, M.R. Ganjali, M. Zare, Application of an Hg2+ selective imprinted polymer as a new modifying agent for the preparation of a novel highly selective and sensitive electrochemical sensor for the determination of ultratrace mercury ions, Anal. Chim. Acta 689 (2011) 52–59.

[17] H.N. Kim, W.X. Ren, J.S. Kim, J. Yoon, Fluorescent and colorimetric sensors for detection of lead, cadmium, and mercury ions, Chem. Soc. Rev. 41 (2012) 3210–3244.

[18] Y.-F. Zhu, Y.-S. Wang, B. Zhou, J.-H. Yu, L.-L. Peng, Y.-Q. Huang, X.-J. Li, S.-H. Chen, X. Tang, X.-F. Wang, A multifunctional fluorescent aptamer probe for highly sensitive and selective detection of cadmium(II), Anal. Bioanal. Chem. 409 (2017) 4951–4958.

[19] M. Caselli, Porphyrin-based electrostatically self-assembled multilayers as fluorescent probes for mercury(II) ions: a study of the adsorption kinetics of metal ions on ultrathin films for sensing applications, RSC Adv. 5 (2015) 1350–1358.

[20] J.-L. He, S.-L. Zhu, P. Wu, P.-P. Li, T. Li, Z. Cao, Enzymatic cascade based fluorescent DNAzyme machines for the ultrasensitive detection of Cu(II) ions, Biosens. Bioelectron. 60 (2014) 112–117.

[21] S. Cai, Y. Lu, S. He, F. Wei, L. Zhao, X. Zeng, A highly sensitive and selective turn-on fluorescent chemosensor for palladium based on a phosphine-rhodamine conjugate, Chem. Commun. 49 (2013) 822–824.

[22] J. Yang, Z. Wang, Y. Li, Q. Zhuang, W. Zhao, J. Gu, Porphyrinic MOFs for reversible fluorescent and colorimetric sensing of mercury(II) ions in aqueous phase, RSC Adv. 6 (2016) 69807–69814.

[23] H. Xu, R. Miao, Z. Fang, X. Zhong, Quantum dot-based turn-on fluorescent probe for detection of zinc and cadmium ions in aqueous media, Anal. Chim. Acta 687 (2011) 82–88.

[24] F. Wu, H. Su, K. Wang, W.-K. Wong, X. Zhu, Facile synthesis of N-rich carbon quantum dots from porphyrins as efficient probes for bioimaging and biosensing in living cells, Int. J. Nanomed. 12 (2017) 7375–7391.

[25] E. Jang, S. Jun, H. Jang, J. Lim, B. Kim, Y. Kim, White-light-emitting diodes with quantum dot color converters for display backlights, Adv. Mater. 22 (2010) 3076–3080.

[26] V. Renugopalakrishnan, B. Barbiellini, C. King, M. Molinari, K. Mochalov, A. Sukhanova, I. Nabiev, P. Fojan, H.L. Tuller, M. Chin, et al. Engineering a robust photovoltaic device with quantum dots and bacteriorhodopsin, J. Phys. Chem. C 118 (2014) 16710–16717.

[27] E. Mohamed Ali, et al. Ultrasensitive Pb^{2+} detection by glutathione-capped quantum dots, Anal. Chem. 79 (24) (2007) 9452–9458.

[28] Z.S. Qian, X.Y. Shan, L.J. Chai, J.J. Ma, J.R. Chen, H. Feng, A universal fluorescence sensing strategy based on biocompatible graphene quantum dots and graphene oxide for the detection of DNA, Nanoscale 6 (2014) 5671–5674.

[29] X. Ran, H. Sun, F. Pu, J. Ren, X. Qu, Ag nanoparticle-decorated graphene quantum dots for label-free, rapid and sensitive detection of Ag+ and biothiols, Chem. Commun. 49 (2013) 1079–1081.

[30] I. Al-Ogaidi, H. Gou, Z.P. Aguilar, S. Guo, A.K. Melconian, A.K. Al-kazaz, F. Meng, N. Wu, Detection of the ovarian cancer biomarker CA-125 using chemiluminescence resonance energy transfer to graphene quantum dots, Chem. Commun. 50 (2014) 1344–1346.

[31] S. Zhu, J. Zhang, C. Qiao, S. Tang, Y. Li, W. Yuan, B. Li, L. Tian, F. Liu, R. Hu, H. Gao, H. Wei, H. Zhang, H. Sun, B. Yang, Strongly green-photoluminescent graphene quantum dots for bioimaging applications, Chem. Commun. 47 (2011) 6858–6860.

[32] Q. Liu, B.D. Guo, Z.Y. Rao, B.H. Zhang, J.R. Gong, Strong two-photon-induced fluorescence from photostable, biocompatible nitrogen-doped graphene quantum dots for cellular and deep-tissue imaging, Nano Lett. 13 (2013) 2436–2441.

[33] X.M. Li, S.P. Lau, L.B. Tang, R.B. Ji, P.Z. Yang, Multicolour light emission from chlorine-doped graphene quantum dots, J. Mater. Chem. C 1 (2013) 7308–7313.

[34] L.L. Li, J. Ji, R. Fei, C.Z. Wang, Q. Lu, J.R. Zhang, L.P. Jiang, J.J. Zhu, A facile microwave avenue to electrochemiluminescent two-color graphene quantum dots, Adv. Funct. Mater. 22 (2012) 2971–2979.

[35] L.B. Tang, R.B. Ji, X.M. Li, K.S. Teng, S.P. Lau, Energy-level structure of nitrogen-doped graphene quantum dots, J. Mater. Chem. C 1 (2013) 4908–4915.

[36] R. Balog, B. Jorgensen, L. Nilsson, M. Andersen, E. Rienks, M. Bianchi, Bandgap opening in graphene induced by patterned hydrogen adsorption, Nat. Mater. 9 (2010) 315–319.

[37] K. Gopalakrishnan, K.S. Subrahmanyam, P. Kumar, A. Govindaraj, C.N.R. Rao, Reversible chemical storage of halogens in few-layer graphene, RSC Adv. 2 (2012) 1605–1608.

[38] R.R. Nair, W. Ren, R. Jalil, I. Riaz, V.G. Kravets, L. Britnell, Fluorographene: a two-dimensional counterpart of Teflon, Small 6 (2010) 2877–2884.

[39] X.M. Li, S.P. Lau, L.B. Tang, R.B. Ji, P.Z. Yang, Sulphur doping: a facile approach to tune the electronic structure and optical properties of graphene quantum dots, Nanoscale 6 (2014) 5323–5328.

[40] F. Li, C.J. Liu, J. Yang, Z. Wang, W.G. Liu, F. Tian, Mg/N double doping strategy to fabricate extremely high luminescent carbon dots for bioimaging, RSC Adv. 4 (2014) 3201–3205.

[41] Y. Wang, Y.Y. Shao, D.W. Matson, J.H. Li, Y.H. Lin, Nitrogen-doped graphene and its application in electrochemical biosensing, ACS Nano 4 (2010) 1790–1798.

[42] M. Li, W.B. Wu, W.C. Ren, H.M. Cheng, N.J. Tang, W. Zhong, Synthesis and upconversion luminescence of N-doped graphene quantum dots, Appl. Phy. Lett. 101 (2012) 103107.

[43] C.F. Hu, Y.L. Liu, Y.H. Yang, J.H. Cui, Z.R. Huang, Y.L. Wang, One-step preparation of nitrogen-doped graphene quantum dots from oxidized debris of graphene oxide, J. Mater. Chem. B 1 (2013) 39–42.

[44] I.S. Amiinu, J. Zhang, Z. Kou, X. Liu, O.K. Asare, H. Zhou, K. Cheng, H. Zhang, L. Mai, M. Pan, S. Mu, Self-organized 3D porous graphene dual-doped with biomass-sponsored nitrogen and sulfur for oxygen reduction and evolution, ACS Appl. Mater. Interfaces 8 (2016) 29408–29418.

[45] R. Li, Z. Wei, X. Gou, Nitrogen and phosphorus dual-doped graphene/carbon nanosheets as bifunctional electrocatalysts for oxygen reduction and evolution, ACS Catal. 5 (2015) 4133–4142.

[46] S. Ju, W.P. Kopcha, F. Papadimitrakopoulos, Brightly fluorescent single-walled carbon nanotubes via an oxygen-excluding surfactant organization, Science 323 (2009) 1319–1323.

[47] S.J. Yu, M.W. Kang, H.C. Chang, K.M. Chen, Y.C. Yu, Bright fluorescent nanodiamonds: no photobleaching and low cytotoxicity, J. Am. Chem. Soc. 127 (2005) 17604–17605.

[48] S.N. Baker, G.A. Baker, Luminescent carbon nanodots: emergent nanolights, Angew. Chem. Int. Ed. Engl. 49 (2010) 6726–6744.

[49] J. Gu, X. Zhang, A. Pang, J. Yang, Facile synthesis and photoluminescence characteristics of blue-emitting nitrogen-doped graphene quantum dots, Nanotechnology 27 (16) (2016) 165704.

Machine learning analysis of topic modeling re-ranking of clinical records

Vijayalakshmi Kakulapati[a], Sheri Mahender Reddy[a], B. Sri Sai Deepthi[b], João Manuel R.S. Tavares[c]

[a]Sreenidhi Institute of Science and Technology, Hyderabad, India; [b]Mamatha Medical College, Khammam, India; [c]Instituto de Ciência e Inovação em Engenharia Mecânica e Engenharia Industrial, Departamento de Engenharia Mecânica, Faculdade de Engenharia, Universidade do Porto, Porto, Portugal

1 Introduction

Today, electronic medical records (EMRs) are created by many technologies like sensors, wearable, and devices. The integration of wearable devices and medical records can be used to study a variety of physical conditions, for example, in fitness. The combination of electronic health record (EHR) with big data, for instance, current medical records into the EMRs together with inherent data, is very promising. This type of combination of EHRs can supply important, complete, and consistent basis of data for medical studies [1]. The main purpose of wearable devices is to generate data without human intervention by zero attempts need from enduring [2]. On the contrary, the manual creation of health records of patients visiting a hospital is much more time consuming, especially when testing different parameters like level of glucose in blood and heartbeat rate. Whereas wearable devices are efficient and precise in time, usually, such devices can verify all the acquired features and analyze them directly, unlike manual methods. In this chapter, the discharge sheets are generated by taken into account EHR details and analyzing them based on topic modeling technology. The topic models are used to detect behavior patterns in EMRs, which are produced by a health monitoring system.

The medical reports of patients provide information regarding growing data for managing patient information and predicting trends in diseases. Healthcare service providers face a major challenge regarding patients who are suffering from multiple problems with inefficient diagnosis, which leads to frequent and increasing visits to hospitals. Therefore, the discover of symptoms related to health conditions is increasingly interesting since it can help to obtain improved predictions for hospitalization, disease, or death. In this regard, Moumita Bhattacharya et al. [3] proposed the EMRs topic modeling to analyze the symptoms and patient behavior.

Summarization in text mining is one of the major challenges. Because of this summarization, researchers give information to stakeholders and developed several real-time applications. The huge number of documents is converted into a decreased and compacted in summarization indicates the summary of the document collections. The document summarization gives better knowledge about the overall content of the dataset. This summarization reduces the physician time consumption without reading of the entire patient report. Generally, a function of converting entire document information to small chunks is called as document summarization. These chunks of information hold the entire description of the document collection.

Text summarizer carried out by the algorithm is derived from the text summarization task. These are classified into two ways: one is single document and another one is multi-document. The first type, which is a single document, is summarized in the summary of the document, whereas, in second one, collection documents are summarized in the summary of the document, which gives the total knowledge of the various documents.

One of the popular statistical-based approaches for the allocation of items in large corpus into subsets which are semantically-meaningful and used on textual corpus is LDA topic modeling. Documents are arbitrary combinations over topics in the dataset, which makes logic between topics which is exceptional as regards to a particular topic. For example, a news article on the President of the United States moves toward healthcare. The topics in the news would be reasonable to allocate like President, the United States, health assurance, and political opinions, though it is to confer the medicinal service.

The dataset contains documents which are a collection of a related number of topics; these topics are associated with a diversity of phrases which shows every document is the consequence of a combination of probabilistic samples: possible topics distribution and selected topic possible word listing—one of the main advantages of LDA than probabilistic latent semantic analysis (PLSA) and latent semantic indexing (LSI) topic modeling techniques. LDA is a generative model which employs to split the text into the topic to document on outside the dataset. For instance, LDA groups news articles into classes like sports, entertainment, and politics, to the potential use of the fitted model to facilitate the classification of recently-circulated news. This facility is away from the scope of approaches like LSI. The number of parameters to an approximation for LDA model dimensions with the number of topics is much lower number, which makes LDA is apt to effective with huge datasets. LDA is to model documents as occurring from several topics, where each topic is described to be an allocation over a fixed glossary of words. Each document is a collection of topics and shows these topics with diverse parts because documents in a dataset are apt to be heterogeneous, merging a subset of main themes that filter through the group as a whole.

Nowadays, researchers are concentrating on summarization techniques for the document. Numerous methods are developing to digest by retrieving the significant topics from particular corpora. For analyzing the unstructured text is utilizing probabilistic topic models, and provides the latent to be incorporating into patient medical record summarization. A patient medical record contains metadata about the patient's diagnosis history and multiple topic concepts that can be precious for exactly understanding the document. The unified model [4] is utilizing for free-text medical reports which incorporate appropriate patient and data at document level and identify multiword in medical documents.

Clinical reports contain information regarding the patient is accumulating as the free text in the general practitioner's medical documents. These reports give medical description can be a computationally challenging task to make understandable by the inconsistency in physicians writing styles, disparities in their observation, and the intrinsic linguistic. However, a clinical report provides case-based reasoning (CBR) [5] and automatic summarization [6] data for medical applications. Topic modeling of documents provides indexing large, unstructured data with conditional semantics [7]. These methods show potential results due to these basic methods that have not integrated further progression in the field of topic modeling. Improvement of advances in topic modeling methods shows varied medical data and potential structure to release the data in medical documents. Medical document processing and summarization of a database is a difficult undertaking task and in the Big Data era, where data is more and more, algorithms for summarizing the large clinical reports are required.

2   Related work

In [8], topic models explain about how to produce the concept from a prescription combination. In the same way, traditional Chinese medicine utilized the interactions between herbs to retrieve symptoms and analyzed [9] variants of LDA. Though, topic modeling using in clinical documents analyzing is a promising field. Topic modeling of unstructured clinical documents is classified and represents clinical reports. By utilizing topic models, the content has been exploring for an association between symptoms and topic adaptations that are topic-concept models [10,11]. Similarly, the investigation of entertaining drug conversations [12] and pertinent to clinical practice, clinical case repossession could be addressed [13]. In this work, we focus on the patient discharge summary report and the involvement of different patient-related information.

Text data contain bag-of-words (BoW) required to be changing to an appropriate format for computerized processing. For BoW, each report develops into a token/word vector. Patient clinical information is retrieved by analyzing EHRs [14]. These EHR data contain empty spaces and need to be preprocessing to utilize in computer-based methods. Using these data could be efficient and effective for the speed and quality of healthcare. In our implementation, we utilize discharge summary reports. Generally, for regular text classification, topic modeling is implemented on the entire dataset in diverse methods. Clinical reports topic modeling provides understandable topics which exist in medical reports. This type of representing reports based on their topic allocations is additional dense than representation of BoW and can be improving in-process documents than raw text in successive computerized processes.

For generating topic models of discharge summary reports, we utilize LDA due to the probabilistic system for clinical documents and its toughness to overfitting. LDA believes that medical reports contain underlying topics and every topic classified by an allocation transversely words [15]. LDA is utilizing for a large variety of healthiness and clinical applications for predicting textual data [16], learning appropriate medical models and arrangements in clinical records [17], identifying prototypes of medical events in brain cancer patients [18], and examining the results [19]. The pattern contains information about enlightening the formation, semantics, and dynamics. These patterns give physicians with precise information which is utilized to guide better treatment actions of each patient. For finding treatment behavior of patients, LDA utilized these patterns [20] to predict medical classifying patterns, and to form diverse diagnosis activities [21] and pattern time stamps [22]. To determine enduring transience customized by LDA [23] and also identifying the knowledge based on characteristics of the patient and modeling disease [24]. Better performance than LDA for managing issues related to redundancy in clinical report using redundancy-aware LDA [25].

Generating summaries from the huge collection of documents is the intention of MapReduce construction-based summarization technique. Implementation results evaluation time for summarizing the huge collection of documents is significantly decreased utilizing this framework and also offers scalability for accepting huge document assortments for summarizing, which is a trendier programming model for processing huge data. By using MapReduce, this provides several benefits in maintaining a huge amount of data, for example, scalability, flexibility, fault tolerance, and several benefits. Nowadays, many researchers [26–32] are presented in several works in the aspect of big data and processing of the huge amount of data. It is extensively utilized for processing and handling the huge amount of data in a disseminated cluster, which has been utilized for several domains, for example, text clustering, access log investigation, creating search catalogs, and diverse data analytical functions. The MapReduce framework [33] is to execute clustering on the huge amount of data by utilizing customized K-means clustering algorithm.

The MapReduce framework is effectively employed for several document processing tasks for dealing with large text that are the complicated tasks in the knowledge discovery process. In text analytics, summarizing the huge amount of text set is a motivating and challenging crisis. Many researchers propose for dealing large text for automatic text summarization (ATS) [34,35]. Utilizing prosodic elements and enhance lexical element technique is proposed [36] for gathering summarization. An unsupervised technique [37] uses for the regular summarization of source code text, which is employed for code folding, and allocates one to discriminating conceal chunks of code.

Parametric shortest path algorithm utilizing phrase graphs is a multi-sentence compression technique [38] that presents for multi-sentence compression. For creating the required summary, a parametric method of edge weights is utilized. The execution is carried out by utilizing the MPI and framework of MapReduce, which is exhibited by parallel implementation of latent Dirichlet allocation (PLDA) [39] that can be useful to huge, real-world applications and accomplishes superior scalability.

3   Health and medical topic modeling

In text mining, two leaning approaches are there: classification, which is known as supervised, and clustering, which is known as unsupervised. The first approach is to make the unknown formation in labeled datasets, whereas the second one is to identify the patterns in unlabeled data collection. The supervised learning method is the classification, and the unsupervised learning method is clustering. The first approach is to prepare data with labels that are predefined and assign to a new record [40]. Unsupervised learning allocates a set of every record in a data collection based on clustering similarity functions. Topic modeling is the most acceptable clustering techniques for a broad category of applications. Topic modeling deriving every topic is distribution of probability words and reports as probability distribution over topics. In clinical reports mining LDA gives more relevant information than other models.

In large corpus, topic modeling is unsupervised learning, which discovers the contents of a text collection. Techniques utilized latent semantic analysis (LSA) [41], PLSA [42], and LDA [43]. The unknown semantic arrangement of a word-text matrix where the text is rows and words are columns [44] depends on singular value decomposition. The main disadvantage of LSA is every word is delighted as the similar meaning: word polysemes cannot distinguish. The result of this analysis that consists of axes in Euclidean space is not understandable [45].

CBR is a technique implemented from knowledge-based classification in diverse provinces, which utilizes occurrences from prior related cases to resolve the latest crisis. The reason behind CBR is the hypothesis that related cases have analogous solutions [46]. By using CBR in different research problems, including similarity estimation algorithms, catalog methods enhance the effectiveness of retrieval methods, case depiction techniques, and techniques to add the latest cases [47]. CBR main the history of past cases before the individual determined in rules, every case includes a depiction of the case, that is, solution, which is the implicit solution.

CBR used to solve the latest case is that the case matched beside the cases in the case base, and analogous cases are repossessed, which is utilized to imply a solution reprocessed and examined for accomplishment. At last, the most recent case and its solution saved as the segment of a most recent case.

Creating a short, precise, and assured summary of a longer text document is known as text summarization. ATS techniques are required to address a large amount of text data accessible online to assist relevant information retrieval and reducing the user retrieval process. ATS is a text summarization, which is the procedure of generating a small and logical description of a large document.

LDA and LSI are statistical methods, whereas the former one is used for complex probability and later used for simple. LSI is less complex than LDA, and LDA is a considerable extension of LSI. The major weakness of LSI is ambiguity. In LDA, words grouped into topics, which can exist in more than one topic. LDA deals with ambiguity by evaluating a document to two topics and resolving which topic is nearer to the document, transversely all permutations of topics. LDA assists the search engine to establish which documents are most significant to which topics. PLSA is identical to LDA except that the topic allocation is supposed to have sparse Dirichlet prior (SDP). SDPs determine the perception that documents cover only a few topics; these topics utilize only a few words frequently. The results are disambiguation of words and the precise task of documents to topics. The generalization process of PLSA model is LDA.

The probabilistic version of LSA is PLSA where an unseen variable is related to every incidence of a topic in a specific record. Topics are then contingent from the participation in clinical reports. The polysemy problem is solved by PLSA, but it is not considering a completely generative model of reports which is called as overfitting. Multiple factors produce linearly with numerous documents. Topic distribution describing in LDA over a fixed language and every document can display topics with diverse sections. LDA creates the topics in a two-step process for every medical report:

1. Topics are arbitrarily choosing in an allocation.
2. For every topic in the report:
   a. Arbitrarily select a word from the allocation over words.
   b. Arbitrarily select a topic from the consequent language distribution.

The possibility of creating the topic t_j from report r_i can be defined as follows:

$$P\left(T_j \big| r_i; \theta, \varphi\right) = \sum_{k=1}^{K} P\left(Tj \big| zi; \varphi\right) P\left(zk \big| di; \theta d\right)$$

where θ is a model from the distribution of Dirichlet for every text di and Θ model from the distribution of Dirichlet for each word zk. Using different sampling methods like Gibbs sampling [48] and optimization methods [49] is to prepare a topic model in LDA. The efficiency of LDA is better than PLSA for simple data collection as it avoids overfitting and polysemy support. In dissimilarity of PLSA, LDA has also considered a completely creative method for text.

LDA is an extension of PLSA where the topic and word allocations have Dirichlet priors. PLSA supposes that have consistent prior. The term allocations in LDA $p(w|z)$ have Dirichlet preceding with parameter α, and the topic allocations $p(z|d)$ have Dirichlet preceding with parameter β. Empirical experiments in LDA show better PLSA in cases where the number of parameters largely evaluated to the size of the data [50].

The LDA [51] is an effort to get better PLSA by establishing a Dirichlet prior on document-topic allocation. Multinomial distributions of prior association [52] of Dirichlet prior simplify the statistical inference problem. The LDA [53] successfully applied in diverse applications for recognizing topics. Performance of the LDA compared with other models, such as unigram, mixture unigram, and the PLSA in terms of perplexity. In this, they addressed that the LDA demonstrated superior performance and also LDA is not experiencing the severe overfitting crisis, which is related with the PLSA.

MapReduce [54] is a program representation and a related implementation for doing out and creating big corpus with an equivalent, disseminated cluster algorithm. We utilized a novel structure, which is based on MapReduce tools for summarizing the huge document collection. This method is determining to by means of clustering semantic similarity and topic modeling utilizing LDA for the document collection summarization. The main advantage of the proposed framework is observable from the testing and also affords a faster execution of summarizing the huge collection of documents and is an influential tool in analysis of big data.

Conversely, the results retrieved by LDA [55] may not be initiative for understandable format and use. In our proposed model, we implement various topic and keyword re-ranking approaches which help stakeholder's healthier knowledge and utilize the words derived by LDA in the analysis of records. We utilized techniques to process the LDA results depend on a set of conditions that will provide required information for the patient. Our experiment analysis exhibits the effectiveness of the techniques in summarizing patient discharge summary reports.

4 Framework

4.1 Patient reports

"De-identifying discharge summary reports using in this investigation are provided by the i2b2 National Center for Biomedical Computing funded by U54LM008748 and dataset is preparing for the sharing of Challenges in NLP for Clinical Data organizing by Dr. Ozlem Uzuner, i2b2 and SUNY." The dataset contains 390 discharge summaries for different patients. These reports contain details of the patient like patient history, symptoms, patient id, etc. In this dataset, the dataset is collected from the homogeneous set of patients from a medical perspective. For implementation purpose, the dataset is categorized as the training set and test set.

The patient discharge summaries including patient name and all patient-related information, prescriptions, and conditions illustrate the enduring (e.g., heart pain). The patient's discharge summary perceptions in this assignment consist of things linked to a long suffering, which are frequent in medical reports and determine co-references are crucial for the receipt of an overall description of the clinical situation. The discharge summary that consists of different topic, such as the "patient history" and "prescription," describes the patient data in diverse situations. Additionally, the text format is not particular; so numerous names can be present in each entity. For instance, physician names, clinician, doctor, etc., can refer to the similar person in a medical report summary.

4.2 Dataset

This study used i2b2 patient discharge summary report and each report contains the patient-related information and this dataset contains 380 discharge summaries. The summary contains above 4000 topics that indicate the patient-relevant information that is patient id, name, patient history, symptoms, medication, etc. We classified this dataset into training set about 100 records and 200 discharge summaries are test dataset.

The clinical discharge summary dataset also included metadata for each report concerning data about the therapeutic history, with the date, the name of clinician, and prescription status. In the same way, demographic data were related with every discharge summary report, as well as the age of patient, gender, family history, and course. These data were through existing to the representation to utilize as earlier in generating topics precise to the present report.

4.3 Preprocessing

The clinical discharge summary dataset cleaned every medical report to remove irrelevant inconsistencies in the data collection. Subsequent to cleaning, the discharge summary collection consists of 4000 topics in total.

In topic modeling the input data are a document-term matrix, where the tuples equal to text and the attributes to the words. The total number of tuples is corresponding to the dataset size and the total number of attributes to the magnitude of the language. Document mapping to the term occurrences of vector includes tokenizing the text and then handling the tokens, for instance, by translating tokens to lower case, eliminating punctuations, eliminating numbers, stemming, eliminating stop words and the missing terms with a length under a certain minimum.

4.4 LDA model

LDA defines the topic as a distribution of language, where each report demonstrates with diverse proportions. LDA utilizes probabilities and characterizes documents as the combination of topics that categorize words with certain probabilities. Discharge summaries are produced in the subsequent approach:

- According to Poisson distribution, the number of terms in the text.
- According to the distribution of Dirichlet distribution more than a predetermined set of K topics, select a topic combination for the document.
- Create every topic Ti in the report by:
  - A topic selection
  - According to the topic's multinomial distribution, the topic of creating the word itself.

Reporting generative representation, LDA tries to back off the reports to discover a collection of topics created the set.

4.5   The distribution of topics

Patient discharge summary reports topic modeling generates a distribution of topics for every summary report. These topics can be utilized as topic vectors, which correspond to another approach for BoW. In these topic vectors, terms are swapping in every summary document that shows the probability of a precise topic with topics and entries for that report. The topic vector concept is further precise than BoW as the languages for a report generally have thousands of entries, while a topic model usually constructed with a limit of topics.

5   Experiment analysis

For evaluating the accuracy of the model, which is used in machine learning algorithm, a significant measure was used to interpret the outcomes. Supervised classification is the best in this step and is straightforward, for instance, as a class label is known in supervised learning of the data classified, the performance can be evaluated by simply calculating the number of faults. In the topic modeling the condition is not so simple, with LDA utilizing an algorithm to identify logical subgroupings in data. Usually, in evaluation, it should continue with an assessment of homogeneity of the words consisting of the documents in every grouping. In topic modeling [56], it is possible to calculate the topic model from a statistical perception utilizing holdout investigating document assortment. Implementing LDA on document dataset observes the topmost frequent words that can originate in every group. Every document can allocate to a topic, based on the combination of topics. LDA will allocate every document is a set of possibilities analogous to every probable topic.

5.1   Extracting and visualizing topics

5.1.1   Extraction of topics by using latent Dirichlet allocation

The most popular topic modeling technique is LDA, which forms clinical discharge summaries as the combination of hidden topics; these topics are main models existed in the report. Clinical report in the topic model is a probability model, whereas every report congaing a grouping of topics and these topics correspond to the collection of words that is inclined to happen mutually. Φk is every topic represented as the distribution of probability over lexical words. Every topic is representing as a word's vector with the probability. A clinical discharge summary is characterized as an allocation of probability topics.

The LDA topic modeling process depends on a combined distribution of probability between topics unknown and the words observed to collect the words with the probability elevated in every topic by utilizing the posterior distribution. In LDA, the popularly accepted method collapsed Gibbs sampling used in analyzing the results. These methods require several repetitions lead to the cost of computational linearly with multiple clinical reports.

In our clinical discharge summaries, the resulting topics and patterns originate from associating with suitable topics of medical reports. Table 9.1 shows various

Table 9.1 List of symptoms.

S. No.	Symptom 1	Symptom 2	Symptom 3	Symptom 4	Symptom 5	Symptom 6	Symptom 7	Symptom 8	Symptom 9	Symptom 10
1	endometri	aortic	unit	arteri	histori	confirm	date	blood	hemorrhag	overrid
2	recent	cord	per	diseas	medic	ultim	follow	status	magnet	amiodaron
3	gallbladd	cathet	time	coronari	left	around	summari	cell	tomographi	elev
4	pelvic	aneurysm	hospit	cardiac	admiss	lenni	diagnosi	white	reson	interact
5	duct	spinal	given	underw	normal	breutzoln	report	chest	side	hcl
6	nonfoc	everi	care	left	hospit	degen	procedur	increas	gait	start

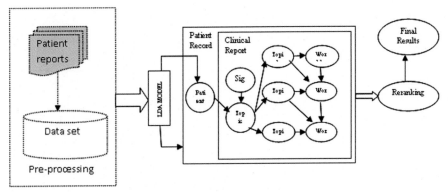

FIGURE 9.1 Re-ranking framework for patient clinical records.

symptoms obtained by the model and Fig. 9.1 shows the frequency of symptoms, and their probability is shown in Table 9.2 from the clinical discharge summary dataset. Topics obtained by learning across all patients; generally, patients exhibit a subset of all potential topics. Medical report dataset where similar words are employed across summary records is too complex because there are several unique words connected to the total number of words.

Our dataset contains above 4000 topics. Some of them are as follows:

"Discharge","histori","medic","admiss","hospit","date","pain","status,"normal" ,"blood","time","show","follow","report","diagnosi","present", "cours", "examin", "admit", "year", "summari", "diseas", "past", "sign", "care", "bilater" and many more.

5.1.2 *Categorization of disease symptom categorization*

5.1.2.1 Without symptom-interdependency models

In this category, each disease treats different independent symptoms. This type is generally using in vector space models by the orthogonality hypothesis of symptom vectors by an independency assumption of symptom variables.

5.1.2.2 With immanent symptom interdependency models

This type of representation allows interdependencies between symptoms, whereas the degree of the interdependency between two symptoms is defining the model itself. These models are straightforward or not directly derived from the co-occurrence of symptoms in the entire set of clinical reports.

5.1.2.3 With transcendent term interdependency models

This type of representation allows interdependencies between symptoms. These models do not assert how the interdependency between the two symptoms is derived.

The generative model in LDA is summarized as follows:

1. For every topic, choose what words are probable.

Table 9.2 Probability of symptoms for disease by using LDA and ranking.

Patient id	Symptom 1	Symptom 2	Symptom 3	Symptom 4	Symptom 5	Symptom 6	Symptom 7	Symptom 8	Symptom 9	Symptom 10
1	0.0093	0.0148	0.0391	0.0670	0.0335	0.0111	0.2290	0.0689	0.0130	0.5139
2	0.0760	0.0869	0.0652	0.0543	0.0543	0.0543	0.4347	0.0652	0.0543	0.0543
3	0.0467	0.0607	0.0560	0.0747	0.2476	0.0747	0.0607	0.1915	0.1168	0.0700
4	0.0311	0.0249	0.0623	0.3956	0.2585	0.0373	0.0716	0.0436	0.0467	0.0280
5	0.0701	0.0491	0.0596	0.0877	0.3929	0.0421	0.1017	0.0491	0.1228	0.0245
6	0.0436	0.0187	0.0769	0.0852	0.1559	0.0727	0.0415	0.4137	0.0602	0.0311
7	0.0331	0.0165	0.0900	0.2559	0.2417	0.1042	0.0687	0.0781	0.0781	0.0331
8	0.0537	0.0950	0.0619	0.0785	0.3181	0.0289	0.2107	0.0619	0.0702	0.0206
9	0.1710	0.0427	0.0690	0.0230	0.3223	0.0263	0.1282	0.1085	0.0789	0.0296
10	0.0370	0.0679	0.0370	0.0370	0.1358	0.0370	0.4814	0.0617	0.0679	0.0370
11	0.0914	0.0242	0.0471	0.06	0.2428	0.0171	0.0328	0.4571	0.0128	0.0142
12	0.0921	0.0401	0.0141	0.0330	0.3120	0.0543	0.0378	0.3617	0.0236	0.0307
13	0.0555	0.0666	0.1	0.0666	0.0777	0.0888	0.3333	0.0555	0.0888	0.0666
14	0.0438	0.0350	0.0438	0.0657	0.5	0.0438	0.0877	0.0701	0.0526	0.0570
15	0.0193	0.0502	0.0618	0.0541	0.1934	0.2978	0.1005	0.1934	0.0174	0.0116
16	0.0156	0.0254	0.0627	0.4980	0.1941	0.0137	0.0411	0.0980	0.0235	0.0274
17	0.0652	0.0380	0.0326	0.0434	0.2934	0.0380	0.0815	0.2771	0.0652	0.0652
18	0.0628	0.0571	0.0457	0.0342	0.04	0.04	0.6	0.04	0.04	0.04
19	0.0707	0.0530	0.0619	0.0442	0.0796	0.0442	0.4778	0.0442	0.0619	0.0619
20	0.0330	0.0301	0.1149	0.2011	0.1997	0.0186	0.091	0.2068	0.0833	0.0201
21	0.0341	0.0255	0.0447	0.0234	0.3411	0.0362	0.2409	0.1194	0.0298	0.1044
22	0.1351	0.0210	0.0510	0.0750	0.2822	0.0690	0.1351	0.1921	0.0210	0.0180
23	0.0725	0.0483	0.0403	0.0483	0.0645	0.0887	0.4838	0.0564	0.0483	0.0483
24	0.1209	0.0132	0.0491	0.0132	0.3686	0.0453	0.0567	0.2608	0.0378	0.0340
25	0.0512	0.0427	0.0512	0.0512	0.0598	0.0427	0.4786	0.0427	0.1025	0.0769

26	0.0292	0.0133	0.0937	0.1717	0.3922	0.0389	0.0085	0.1644	0.0499	0.0377
27	0.1975	0.0362	0.0443	0.0443	0.125	0.0322	0.3830	0.0564	0.0403	0.0403
28	0.0211	0.0246	0.0563	0.0387	0.2676	0.0774	0.1021	0.2992	0.0774	0.0352
29	0.0884	0.0619	0.0707	0.0442	0.0619	0.0530	0.4867	0.0442	0.0442	0.0442
30	0.0466	0.0333	0.0433	0.0266	0.49	0.0233	0.04	0.1066	0.1733	0.0166
31	0.0182	0.2145	0.0771	0.2061	0.1725	0.0196	0.0897	0.1556	0.0168	0.0294
32	0.0560	0.0560	0.0467	0.0467	0.0467	0.0654	0.4766	0.0467	0.0654	0.0934
33	0.0221	0.0202	0.5378	0.0904	0.1660	0.0249	0.0784	0.0452	0.0073	0.0073
34	0.0292	0.0439	0.0390	0.5723	0.1707	0.0260	0.0341	0.0292	0.0325	0.0227
35	0.1609	0.0218	0.0300	0.0627	0.3997	0.0163	0.0354	0.2482	0.0095	0.0150
36	0.0820	0.0597	0.0522	0.0522	0.1119	0.0522	0.4104	0.0597	0.0597	0.0597
37	0.0217	0.1959	0.1010	0.1306	0.2363	0.0233	0.0217	0.2270	0.0233	0.0186
38	0.0659	0.0494	0.0549	0.0329	0.1703	0.0714	0.3736	0.0604	0.0604	0.0604
39	0.0555	0.0833	0.0925	0.0462	0.0833	0.0740	0.3981	0.0648	0.0555	0.0462
40	0.0160	0.0140	0.0722	0.1124	0.3293	0.0381	0.0381	0.1726	0.1847	0.0220
41	0.0933	0.03	0.07	0.0766	0.3666	0.11	0.1	0.09	0.0333	0.03
42	0.0342	0.0868	0.0473	0.1105	0.3289	0.0552	0.0789	0.1131	0.0947	0.05
43	0.0515	0.0515	0.0309	0.0360	0.0670	0.0618	0.5670	0.0515	0.0463	0.0360
44	0.0495	0.0965	0.0693	0.2202	0.299	0.0247	0.0643	0.1311	0.0198	0.0247
45	0.0512	0.1002	0.0645	0.1603	0.2516	0.0222	0.0489	0.2293	0.0289	0.0423
46	0.1648	0.0358	0.0609	0.0573	0.3405	0.0430	0.1362	0.0752	0.0322	0.0537
47	0.0124	0.0651	0.1914	0.1511	0.2371	0.0208	0.0166	0.2621	0.0249	0.0180
48	0.0427	0.0539	0.0408	0.0241	0.2918	0.0576	0.0464	0.3680	0.0390	0.0353
49	0.0351	0.0351	0.1022	0.3738	0.2108	0.0383	0.0702	0.0543	0.0255	0.0543
50	0.1076	0.0311	0.0708	0.0396	0.2407	0.0226	0.0368	0.3937	0.0226	0.0339
51	0.0195	0.0160	0.0409	0.0587	0.4750	0.0177	0.0284	0.0498	0.2704	0.0231

2. For every clinical discharge summary report,
   a. Choose what percentage of topics supposed to be in the report.
   b. For every term,
   - Selecting a topic
   - Specified this topic, decide a likely word (created in step 1)

   The probabilistic generative process described as:

1. For every topic k, illustrate an allocation over terms.
2. For every report d,
   a. Illustrate a vector of topic percentages.
   b. For every term,
   - Illustrate a topic assignment
   - Illustrate a term

5.2 Re-ranking of keyword

An association among a set of items, for any two items with probable relations, item one is either "ranked higher than," "ranked lower than," or "ranked equal to" the second item, which is called as a weak order or total preorder of items. None of the items can have a similar ranking. For instance, Google search engine can rank the pages it locates according to relevance information, which is making possible for the user quickly to select the pages according to their wish. Re-ranking is to enhance the precision of retrieval documents. The re-ranking provides more relevant information with higher ranking to the users. After ranking consequences are returning, the user can prefer information of importance as the seed information and apply the re-ranking by which documents re-rank based on similarity measures.

In this work, topic and keyword re-ranking techniques to improve the LDA amount produced for more efficient human consumption. First, illustrate re-rank topic keywords derived from LDA because these keywords order directly influences the semantics and as a result the topic importance. The topic keywords order by LDA cannot be the model for stakeholders to be aware of the topic semantics. For instance, when LDA applied to clinical discharge summary records, common diseases such as diabetes, cancer, heart issues, fever, etc., are generally ranked elevated in numerous topics due to their relevance in all topics. These words are not made use of patients identify knowledgeable topics as all of these are not related to them. For providing better information, the topic keywords derived from the LDA to filter the topic definitions by implementing re-ranking technique.

5.3 Re-ranking of topics

For the randomly ordered derived topic by the LDA, those may not be equally important to the patient. The order of topics, those are more useful and important shown first. Generally, the meaning of importance may be different from one patient to another. For instance, a patient may desire to see the most important symptoms, which cover several summary reports. In this situation, the rank of a symptom would be elevated, because it refers summary report content in the dataset. On the contrary, a patient may be concerned with a group of distinct symptoms that contain the smallest content be related to one another. In such situation, rank symptom depends on their uniqueness in content. Subsequently, it illustrates a small number of independent application symptoms re-ranking techniques that divide the topic-based ranks on diverse ranking conditions.

5.4   Clinical reports re-ranking

In the clinical report re-ranking, the rule [57] is about to rank the symptoms of the patient with the highest probability in the medical reports, which is completing by replacement ranking to redefine topics in discharge summary reports (Figs. 9.2 and 9.3).

*1. Algorithm: Ranking (***clinical reports result set CRRS***)*
Input: **clinical reports result set CRRS**.
Output: Arranging the Result List with Ranking r.
do
if (CRRS i >CRRS j) then
Swap (Ii,Ij)
else
Return CRRS I with ranking Order
Until (no more Items in CRRS)

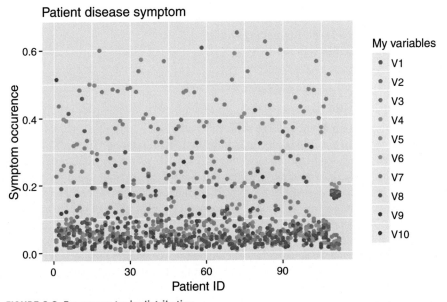

FIGURE 9.2 Frequency topic distribution.

Patient disease plot

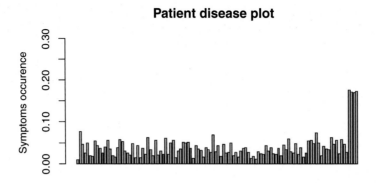

Patient data

FIGURE 9.3 Frequency topic distribution after re-ranking.

In Table 9.1, list of symptoms of discharge summary sheets is presented. All these symptoms are described in Section 5.1.1 (Tables 9.3 and 9.4).

2. Algorithm: Re-ranking (Ranked clinical reports result set RCRRS)
Input: **Ranked clinical reports result set CRRS**
Output: Ordered Result List with Re-Ranking r.
CRD < --GetClinical Report data (q, r, s);
do
if (CRD = True && RCRRS i > RCRRS j) then
Swap (Ii, Ij)
else
Return RCCRS I with Re-ranking Order
Until (no more Items in RCRRS)

6 Conclusion

By incorporating patient discharge summary metadata, over and above, in order to capturing topics in the clinical document, the topic representation of medical reports is improved. The integrate topic modeling of LDA allows the concept, test, and disease studies using discriminating words which are unclear using the BoW method. Common unsupervised methods for topic modeling can determine hidden formation in huge datasets of unstructured medical records. The integration of patient and medical report data generates more knowledge about the prior topics included in a text. Our implementation results of re-ranking technique indicated conditions grouped as topics. The performance achieved by our technique in exhibiting the recognized topics is promising and can be useful in more reliable clinical decision-making, since all the available data are used to identify related symptoms that can be used for facilitating clinical diagnosis with the patient's condition.

Table 9.3   List of symptoms after re-ranking.

S. No.	Symptom 1	Symptom 2	Symptom 3	Symptom 4	Symptom 5	Symptom 6	Symptom 7	Symptom 8	Symptom 9	Symptom 10
1.	duct	aneurysm	hospit	cardiac	admiss	lenni	diagnosi	white	reson	Interact
2.	endometri	aortic	unit	arteri	histori	confirm	date	blood	hemorrhag	Overrid
3.	gallbladd	cathet	time	coronari	left	around	summari	cell	tomographi	Elev
4.	nonfoc	cord	per	diseas	medic	ultim	follow	status	magnet	amiodaron
5.	pelvic	everi	care	left	hospit	degen	procedur	increas	gait	Start
6.	recent	spinal	given	underw	normal	breutzoln	report	chest	side	Hcl

Table 9.4 Probability of symptoms for diseases after re-ranking.

Patient id	Symptom 1	Symptom 2	Symptom 3	Symptom 4	Symptom 5	Symptom 6	Symptom 7	Symptom 8	Symptom 9	Symptom 10
1	0.0149	0.0093	0.0391	0.0670	0.0335	0.0112	0.2291	0.0689	0.0130	0.5140
2	0.0870	0.0761	0.0652	0.0543	0.0543	0.0543	0.4348	0.0652	0.0543	0.0543
3	0.0607	0.0467	0.0561	0.0748	0.2477	0.0748	0.0607	0.1916	0.1168	0.0701
4	0.0312	0.0249	0.0623	0.3956	0.2586	0.0374	0.0717	0.0436	0.0467	0.0280
5	0.0702	0.0491	0.0596	0.0877	0.3930	0.0421	0.1018	0.0491	0.1228	0.0246
6	0.0437	0.0187	0.0769	0.0852	0.1559	0.0728	0.0416	0.4137	0.0603	0.0312
7	0.0332	0.0166	0.0900	0.2559	0.2417	0.1043	0.0687	0.0782	0.0782	0.0332
8	0.0950	0.0537	0.0620	0.0785	0.3182	0.0289	0.2107	0.0620	0.0702	0.0207
9	0.1711	0.0428	0.0691	0.0230	0.3224	0.0263	0.1283	0.1086	0.0789	0.0296
10	0.0679	0.0370	0.0370	0.0370	0.1358	0.0370	0.4815	0.0617	0.0679	0.0370
11	0.0914	0.0243	0.0471	0.0600	0.2429	0.0171	0.0329	0.4571	0.0129	0.0143
12	0.0922	0.0402	0.0142	0.0331	0.3121	0.0544	0.0378	0.3617	0.0236	0.0307
13	0.0667	0.0556	0.1000	0.0667	0.0778	0.0889	0.3333	0.0556	0.0889	0.0667
14	0.0439	0.0351	0.0439	0.0658	0.5000	0.0439	0.0877	0.0702	0.0526	0.0570
15	0.0503	0.0193	0.0619	0.0542	0.1934	0.2979	0.1006	0.1934	0.0174	0.0116
16	0.0255	0.0157	0.0627	0.4980	0.1941	0.0137	0.0412	0.0980	0.0235	0.0275
17	0.0652	0.0380	0.0326	0.0435	0.2935	0.0380	0.0815	0.2772	0.0652	0.0652
18	0.0629	0.0571	0.0457	0.0343	0.0400	0.0400	0.6000	0.0400	0.0400	0.0400
19	0.0708	0.0531	0.0619	0.0442	0.0796	0.0442	0.4779	0.0442	0.0619	0.0619
20	0.0330	0.0302	0.1149	0.2011	0.1997	0.0187	0.0920	0.2069	0.0833	0.0201
21	0.0341	0.0256	0.0448	0.0235	0.3412	0.0362	0.2409	0.1194	0.0299	0.1045
22	0.1351	0.0210	0.0511	0.0751	0.2823	0.0691	0.1351	0.1922	0.0210	0.0180
23	0.0726	0.0484	0.0403	0.0484	0.0645	0.0887	0.4839	0.0565	0.0484	0.0484
24	0.1210	0.0132	0.0491	0.0132	0.3686	0.0454	0.0567	0.2609	0.0378	0.0340
25	0.0513	0.0427	0.0513	0.0513	0.0598	0.0427	0.4786	0.0427	0.1026	0.0769
26	0.0292	0.0134	0.0938	0.1717	0.3922	0.0390	0.0085	0.1644	0.0499	0.0378
27	0.1976	0.0363	0.0444	0.0444	0.1250	0.0323	0.3831	0.0565	0.0403	0.0403

28	0.0352	0.0775	0.2993	0.1021	0.0775	0.2676	0.0387	0.0563	0.0211	0.0246
29	0.0442	0.0442	0.0442	0.4867	0.0531	0.0619	0.0442	0.0708	0.0619	0.0885
30	0.0167	0.1733	0.1067	0.0400	0.0233	0.4900	0.0267	0.0433	0.0333	0.0467
31	0.0295	0.0168	0.1557	0.0898	0.0196	0.1725	0.2062	0.0771	0.0182	0.2146
32	0.0935	0.0654	0.0467	0.4766	0.0654	0.0467	0.0467	0.0467	0.0561	0.0561
33	0.0074	0.0074	0.0452	0.0784	0.0249	0.1661	0.0904	0.5378	0.0203	0.0221
34	0.0228	0.0325	0.0293	0.0341	0.0260	0.1707	0.5724	0.0390	0.0293	0.0439
35	0.0150	0.0095	0.2483	0.0355	0.0164	0.3997	0.0628	0.0300	0.0218	0.1610
36	0.0597	0.0597	0.0597	0.4104	0.0522	0.1119	0.0522	0.0522	0.0597	0.0821
37	0.0187	0.0233	0.2271	0.0218	0.0233	0.2364	0.1306	0.1011	0.0218	0.1960
38	0.0604	0.0604	0.0604	0.3736	0.0714	0.1703	0.0330	0.0549	0.0495	0.0659
39	0.0463	0.0556	0.0648	0.3981	0.0741	0.0833	0.0463	0.0926	0.0556	0.0833
40	0.0221	0.1847	0.1727	0.0382	0.0382	0.3293	0.1124	0.0723	0.0141	0.0161
41	0.0300	0.0333	0.0900	0.1000	0.1100	0.3667	0.0767	0.0700	0.0300	0.0933
42	0.0500	0.0947	0.1132	0.0789	0.0553	0.3289	0.1105	0.0474	0.0342	0.0868
43	0.0361	0.0464	0.0515	0.5670	0.0619	0.0670	0.0361	0.0309	0.0515	0.0515
44	0.0248	0.0198	0.1312	0.0644	0.0248	0.2995	0.2203	0.0693	0.0495	0.0965
45	0.0423	0.0290	0.2294	0.0490	0.0223	0.2517	0.1604	0.0646	0.0512	0.1002
46	0.0538	0.0323	0.0753	0.1362	0.0430	0.3405	0.0573	0.0609	0.0358	0.1649
47	0.0180	0.0250	0.2621	0.0166	0.0208	0.2372	0.1512	0.1914	0.0125	0.0652
48	0.0353	0.0390	0.3680	0.0465	0.0576	0.2918	0.0242	0.0409	0.0428	0.0539
49	0.0543	0.0256	0.0543	0.0703	0.0383	0.2109	0.3738	0.1022	0.0351	0.0351
50	0.0340	0.0227	0.3938	0.0368	0.0227	0.2408	0.0397	0.0708	0.0312	0.1076
51	0.0231	0.2705	0.0498	0.0285	0.0178	0.4751	0.0587	0.0409	0.0160	0.0196
52	0.0191	0.0381	0.0817	0.0926	0.0518	0.4796	0.0845	0.0708	0.0381	0.0436
53	0.0442	0.1105	0.2044	0.1934	0.0773	0.1271	0.0552	0.1050	0.0331	0.0497
54	0.0476	0.0186	0.1139	0.0725	0.0166	0.2816	0.0352	0.0911	0.0269	0.2961
55	0.0485	0.0485	0.0583	0.4757	0.0777	0.0583	0.0485	0.0485	0.0680	0.0680
56	0.0921	0.0151	0.3250	0.0268	0.0201	0.2211	0.0151	0.1759	0.0285	0.0804
57	0.0650	0.1155	0.0939	0.0939	0.0397	0.3899	0.0361	0.0614	0.0433	0.0614
58	0.0097	0.0402	0.6080	0.0208	0.0111	0.1953	0.0374	0.0416	0.0166	0.0194

(continued)

Table 9.4 Probability of symptoms for diseases after re-ranking.(*Cont.*)

Patient id	Symptom 1	Symptom 2	Symptom 3	Symptom 4	Symptom 5	Symptom 6	Symptom 7	Symptom 8	Symptom 9	Symptom 10
59	0.0468	0.0468	0.0809	0.0851	0.0894	0.0255	0.3745	0.0511	0.0340	0.1660
60	0.0333	0.0250	0.0667	0.0944	0.3861	0.0639	0.0444	0.2389	0.0306	0.0167
61	0.0455	0.0265	0.1326	0.0568	0.2121	0.2083	0.1326	0.0985	0.0682	0.0189
62	0.0537	0.0488	0.0902	0.0463	0.2878	0.0927	0.0488	0.2463	0.0463	0.0390
63	0.0199	0.0179	0.0857	0.2231	0.2689	0.0518	0.0876	0.1135	0.0219	0.1096
64	0.0411	0.0274	0.0457	0.0457	0.0457	0.0685	0.4521	0.2009	0.0274	0.0457
65	0.0172	0.0153	0.0460	0.1667	0.0345	0.0153	0.2414	0.0594	0.0153	0.3889
66	0.0409	0.0297	0.0781	0.1227	0.3271	0.0297	0.1636	0.1152	0.0446	0.0483
67	0.0427	0.0366	0.0488	0.0671	0.0427	0.0427	0.4390	0.1220	0.0549	0.1037
68	0.0426	0.0372	0.0904	0.1117	0.2181	0.0638	0.1383	0.1330	0.0372	0.1277
69	0.0359	0.0265	0.1153	0.0964	0.3062	0.0227	0.0435	0.2836	0.0454	0.0246
70	0.0311	0.0115	0.5121	0.0230	0.1415	0.0219	0.0230	0.1968	0.0253	0.0138
71	0.0191	0.0172	0.0134	0.5897	0.2023	0.0153	0.0305	0.0191	0.0153	0.0782
72	0.0315	0.0102	0.6517	0.0331	0.1308	0.0079	0.0449	0.0646	0.0126	0.0126
73	0.0394	0.0276	0.0709	0.1417	0.3701	0.0512	0.1024	0.0984	0.0630	0.0354
74	0.0433	0.0236	0.0630	0.4016	0.1969	0.0394	0.0984	0.0591	0.0551	0.0197
75	0.0293	0.0220	0.0842	0.4103	0.2051	0.0440	0.0623	0.0733	0.0293	0.0403
76	0.0786	0.0429	0.0571	0.0500	0.0643	0.0500	0.5357	0.0357	0.0500	0.0357
77	0.0460	0.0293	0.0837	0.0460	0.2343	0.1046	0.1213	0.1674	0.1255	0.0418
78	0.0750	0.0375	0.0750	0.2000	0.2063	0.0875	0.0750	0.1313	0.0563	0.0563
79	0.0379	0.0238	0.0498	0.0195	0.1928	0.1647	0.0498	0.4204	0.0260	0.0152
80	0.0218	0.0218	0.1987	0.0611	0.2467	0.0175	0.1245	0.0437	0.2402	0.0240
81	0.0564	0.0359	0.0718	0.0667	0.1487	0.0667	0.1026	0.3231	0.1026	0.0256
82	0.0503	0.0186	0.0521	0.0540	0.3818	0.0242	0.0205	0.1508	0.2235	0.0242
83	0.0451	0.0451	0.0451	0.0451	0.0376	0.0376	0.6241	0.0376	0.0376	0.0451
84	0.0380	0.0326	0.0489	0.0272	0.0598	0.0489	0.5870	0.0326	0.0543	0.0707
85	0.0777	0.0583	0.0583	0.0680	0.0680	0.0583	0.4369	0.0680	0.0583	0.0485

86	0.0315	0.0280	0.0490	0.3916	0.2483	0.0210	0.0979	0.0664	0.0245	0.0420
87	0.0247	0.0247	0.1370	0.2795	0.1945	0.0493	0.0904	0.1260	0.0301	0.0438
88	0.2000	0.0483	0.0552	0.0621	0.2793	0.0310	0.1276	0.1276	0.0310	0.0379
89	0.0565	0.0217	0.0478	0.0522	0.0478	0.0348	0.6000	0.0522	0.0609	0.0261
90	0.0386	0.0386	0.0193	0.0611	0.2990	0.0418	0.0675	0.3569	0.0418	0.0354
91	0.0300	0.0158	0.1609	0.3833	0.2019	0.0205	0.0615	0.0836	0.0284	0.0142
92	0.1986	0.0244	0.0557	0.0348	0.1986	0.0209	0.3449	0.0418	0.0174	0.0627
93	0.0806	0.0538	0.0753	0.0430	0.1505	0.0753	0.1290	0.2849	0.0538	0.0538
94	0.0899	0.0562	0.0562	0.0562	0.0787	0.0787	0.3708	0.0562	0.0787	0.0787
95	0.0685	0.0479	0.0822	0.0342	0.0548	0.1027	0.4589	0.0616	0.0548	0.0342
96	0.0737	0.0737	0.0737	0.0737	0.2737	0.0526	0.1474	0.0842	0.0842	0.0632
97	0.0588	0.0490	0.0588	0.0686	0.0784	0.0490	0.4314	0.0686	0.0686	0.0686
98	0.0463	0.0193	0.0502	0.0463	0.4170	0.0386	0.0927	0.2201	0.0347	0.0347
99	0.0410	0.0410	0.3169	0.0574	0.0738	0.0656	0.1202	0.2077	0.0519	0.0246
100	0.0558	0.0340	0.0728	0.0777	0.4830	0.0388	0.1092	0.0461	0.0461	0.0364
101	0.0463	0.0327	0.0490	0.0381	0.3106	0.0436	0.0736	0.0763	0.3106	0.0191
102	0.0417	0.0625	0.0903	0.0347	0.0972	0.0417	0.4792	0.0556	0.0486	0.0486
103	0.0465	0.0465	0.0388	0.0388	0.0543	0.0775	0.5659	0.0465	0.0465	0.0388
104	0.0659	0.0549	0.0659	0.0549	0.0879	0.0769	0.3956	0.0879	0.0549	0.0549
105	0.0243	0.0243	0.0512	0.3558	0.3693	0.0135	0.0566	0.0674	0.0216	0.0162
106	0.0569	0.0569	0.0925	0.0890	0.4306	0.0285	0.0534	0.0783	0.0427	0.0712
107	0.0602	0.0463	0.0880	0.0278	0.0833	0.0694	0.4537	0.0787	0.0509	0.0417
108	0.0691	0.0259	0.0821	0.0799	0.5270	0.0259	0.0670	0.0734	0.0346	0.0151
109	0.1807	0.1772	0.0432	0.0176	0.0102	0.2015	0.0102	0.0219	0.1681	0.1694
110	0.1809	0.1718	0.0489	0.0304	0.0100	0.1978	0.0130	0.0174	0.1709	0.1590
111	0.1820	0.1701	0.0481	0.0319	0.0114	0.1915	0.0132	0.0255	0.1625	0.1638
112	0.1763	0.1735	0.0393	0.0319	0.0091	0.2033	0.0089	0.0221	0.1687	0.1668

In the future, a hierarchical topic model is going to be developed using fuzzy concepts and dynamic application which will automatically summarize patient medical records. The approach will include the topic identification, concept, and time-oriented views, providing support for multilingual text summarization with the help of MapReduce framework to smooth the progress of different medical records.

References

[1] K. Shameer, M.A. Badgeley, R. Miotto, B.S. Glicksberg, J.W. Morgan, et al. Translational bioinformatics in the era of real-time biomedical, health care, and wellness data streams, Brief. Bioinform. 18 (2016) 1–20.

[2] E. Chiauzzi, C. Rodarte, P. DasMahapatra, Patient-centered activity monitoring in the self-management of chronic health conditions, BMC Med. 1 (2015) 1–6.

[3] M. Bhattacharya et al., Identifying patterns of associated-conditions through topic models of electronic medical records, in: 2016 IEEE International Conference on Bioinformatics and Biomedicine (BIBM), IEEE, Shenzhen, China, 2016.

[4] W. Speier, et al. Using phrases and document metadata to improve topic modeling of clinical reports, J. Biomed. Inform. 61 (2016) 260–266.

[5] C.W. Arnold, S.M. El-Saden, A.A.T. Bui, R. Taira, Clinical case-based retrieval using latent topic analysis, AMIA Annu. Symp. Proc. 2010 (2010) 26–30.

[6] J.C. Feblowitz, A. Wright, H. Singh, L. Samal, D.F. Sittig, Summarization of clinical information: a conceptual model, J. Biomed. Inform. 44 (2011) 688–699, doi: 10.1016/j.jbi.2011.03.008.

[7] D.M. Blei, A.Y. Ng, M.I. Jordan, Latent Dirichlet allocation, J. Mach. Learn. Res. 3 (2012) 993–1022, doi: 10.1162/jmlr.2003.3.4-5.993.

[8] J. Dean, S. Ghemawat, MapReduce: a flexible data processing tool, Commun. ACM 53 (1) (2010) 72–77.

[9] D. Borthakur, The hadoop distributed file system: architecture and design, Hadoop Project Website, pp. 1–14. Available from: https://hadoop.apache.org/docs/r1.2.1/hdfs_design.pdf, 2007.

[10] J. Wiens, J.V. Guttag, E. Horvitz, On the promise of topic models for abstracting complex medical data: a study of patients and their medications, in: NIPS Workshop on Personalized Medicine, Sierra Nevada, 2011.

[11] Z. Jiang, X. Zhou, X. Zhang, S. Chen, Using link topic model to analyze traditional Chinese medicine clinical symptom-herb regularities, in: 2012 IEEE 14th International Conference on e-Health Networking, Applications and Services (Healthcom), IEEE, Beijing, China, 2012, pp. 15–18.

[12] E. Sarioglu, K. Yadav, H.-A. Choi, Topic modeling-based classification of clinical reports, ACL, Sofia, Bulgaria, 2013, p. 67.

[13] M. Bundschus, M. Dejori, S. Yu, V. Tresp, H.-P. Kriegel, Statistical modeling of medical indexing processes for biomedical knowledge information discovery from text, in: Proceedings of the 8th International Workshop on Data Mining in Bioinformatics (BIO-KDD'08), Las Vegas, NV, 2008.

[14] M.J. Paul, M. Dredze, Experimenting with drugs (and topic models): multi-dimensional exploration of recreational drug discussions, in: AAAI 2012 Fall Symposium on Information Retrieval and Knowledge Discovery in Biomedical Text, Arlington, VA, 2012.

[15] C.W. Arnold, S.M. El-Saden, A.A. Bui, R. Taira, Clinical case-based retrieval using latent topic analysis, in: AMIA Annual Symposium Proceedings, vol. 2010, American Medical Informatics Association, 2010, p. 26.

[16] E. Sarioglu et al., Topic modeling based classification of clinical reports, in: Proceedings of the ACL Student Research Workshop, Association for Computational Linguistics, Sofia, Bulgaria, August 4–9, 2013, pp. 67–73.

[17] D.M. Blei, A.Y. Ng, M.I. Jordan, Latent Dirichlet allocation, J. Mach. Learn. Res. 3 (2003) 993–1022.

[18] T. Asou, K. Eguchi, Predicting protein-protein relationships from literature using collapsed variational latent Dirichlet allocation, in: Proceedings of the 2nd International Workshop on Data and Text Mining in Bioinformatics, ACM, New York, 2008, pp. 77–80.

[19] C.W. Arnold, S.M. El-Saden, A.A. Bui, R. Taira, Clinical case-based retrieval using latent topic analysis, in: AMIA Annual Symposium Proceedings, vol. 2010, American Medical Informatics Association, 2010, p. 26.

[20] C. Arnold, W. Speier, A topic model of clinical reports, in: Proceedings of the 35th International ACM SIGIR Conference on Research and Development in Information Retrieval, ACM, New York, 2012, pp. 1031–1032.

[21] J.A. Dawson, C. Kendziorski, Survival supervised latent Dirichlet allocation models for genomic analysis of time-to-event outcomes (2012), arXiv preprint arXiv:1202.5999.

[22] Z. Huang, W. Dong, H. Duan, H. Li, Similarity measure between patient traces for clinical pathway analysis: problem, method, and applications, IEEE J. Biomed. Health Inform. 18 (1) (2014) 4–14.

[23] J.H. Chen, M.K. Goldstein, S.M. Asch, L. Mackey, R.B. Altman, Predicting inpatient clinical order patterns with probabilistic topic models vs. conventional order sets, J. Am. Med. Inform. Assoc. 24 (2016) 136.

[24] G. Defossez, A. Rollet, O. Dameron, P. Ingrand, Temporal representation of care trajectories of cancer patients using data from a regional information system: an application in breast cancer, BMC Med. Inform. Decision Making 14 (1) (2014) 24.

[25] M. Ghassemi, T. Naumann, F. Doshi-Velez, N. Brimmer, R. Joshi, A. Rumshisky, P. Szolovits, Unfolding physiological state: mortality modelling in intensive care units, in: Proceedings of the 20th ACM SIGKDD International Conference on Knowledge Discovery and Data Mining, ACM, New York, 2014, pp. 75–84.

[26] R. Pivovarov, A.J. Perotte, E. Grave, J. Angiolillo, C.H. Wiggins, N. Elhadad, Learning probabilistic phenotypes from heterogeneous her data, J. Biomed. Inform. 58 (2015) 156–165.

[27] R. Cohen, I. Aviram, M. Elhadad, N. Elhadad, Redundancy-aware topic modeling for patient record notes, PloS One 9 (2) (2014) e87555.

[28] L. Steve, The Age of Big Data, Big Data's Impact in the World, New York, USA, 2012, pp. 1–5.

[29] P. Russom, Big data analytics, TDWI Research Report, USA, 2011, pp. 1–38.

[30] A. McAfee, E. Brynjolfsson, Big data: the management revolution, Harv. Bus. Rev. 90 (10) (2012) 60–68.

[31] F. Li, B.C. Ooi, M.T. Özsu, S. Wu, Distributed data management using MapReduce, ACM Comput. Surveys 46 (2013) 1–41.

[32] K. Shim, MapReduce Algorithms for Big Data Analysis: Databases in Networked Information Systems, Springer, Berlin, Heidelberg, Germany, (2013) pp. 44–48.

[33] K. Shim, MapReduce algorithms for big data analysis, framework, Proc. VLDB Endow. 5 (12) (2012) 2016–2017.

[34] K.-H. Lee, Y.-J. Lee, H. Choi, Y.D. Chung, B. Moon, Parallel data processing with MapReduce: a survey, ACM SIGMOD Rec. 40 (4) (2011) 11–20.

[35] H.G. Li, G.Q. Wu, X.G. Hu, J. Zhang, L. Li, X. Wu, K-means clustering with bagging and MapReduce, in: Proceedings of the 2011 44th Hawaii International Conference on IEEE System Sciences (HICSS), IEEE, Kauai/Hawaii, USA, 2011, pp. 1–8.

[36] F. Galgani, P. Compton, A. Hoffmann, Citation based summarisation of legal texts, in: Proceedings of 12th Pacific Rim International Conference on Artificial Intelligence, IEEE, Kuching, Malaysia, 2012, pp. 40–52.

[37] M. Hassel, Evaluation of Automatic Text Summarization, Licentiate Thesis, Stockholm, Sweden, 2004, pp. 1–75.

[38] Q. Hu, X. Zou, Design and implementation of multi-document automatic summarization using MapReduce, Comput. Eng. Appl. 47 (35) (2011) 67–70.

[39] C. Lai, S. Renals, Incorporating lexical and prosodic information at different levels for meeting summarization, in: Proceedings of the 15th Annual Conference of the International Speech Communication Association (INTERSPEECH 2014), ISCA, Singapore, 2014, pp. 1875–1879.

[40] J. Fowkes, R. Ranca, M. Allamanis, M. Lapata, C. Sutton, Autofolding for source code summarization, Comput. Res. Repo. 1403 (4503) (2014) 1–12.

[41] E. Tzouridis, J.A. Nasir, L.U.M.S. Lahore, U. Brefeld, Learning to summarise related sentences, in: The 25th International Conference on Computational Linguistics (COLING'14), ACL, Dublin, Ireland, 2014, pp. 1–12.

[42] T.M. Mitchell, Topic modeling for medical data 38, Machine Learning, Web, 1997.

[43] S. Deerwester, S.T. Dumais, G.W. Furnas, T.K. Landauer, R. Harshman, Indexing by latent semantic analysis, J. Am. Soc. Inform. Sci. 41 (6) (1990) 391–407.

[44] T. Hofmann, Probabilistic latent semantic analysis, in: Uncertainity in Artificial Intelligence, UAI'99, Stockholm, 1999.

[45] D.M. Blei, A.Y. Ng, M.I. Jordan, Latent Dirichlet allocation, J. Mach. Learn. Res. 3 (2003) 993–1022.

[46] T. Hofmann, Unsupervised learning by probabilistic latent semantic analysis, Mach. Learn. 42 (1–2) (2001) 177–196.

[47] T.L. Griffiths, M. Steyvers, D.M. Blei, J.B. Tenenbaum, Integrating topics and syntax, in: NIPS'04 Proceedings of the 17th International Conference on Neural Information Processing Systems, ACM, Vancouver, 2004, pp. 537–544.

[48] A. Aamold, Case-based reasoning: foundation issues, AI Commun. 7 (1994) 39–59.

[49] A. Aamold, E. Plaza, Case-based reasoning: foundation issues, methodological variation, and system approach, AI Commun. 7 (1) (1994) 39–59.

[50] A. Asuncion, M. Welling, P. Smyth, Y.-W. Teh, On smoothing and inference for topic models, UAI '09, Montreal, Quebec, Canada, 2009, pp. 27–34.

[51] N.K. Nagwani, Summarizing large text collection using topic modeling and clustering based on MapReduce framework, J. Big Data 2 (2015) 6, doi: 10.1186/s40537-015-0020-5.

[52] D.M. Blei, A.Y. Ng, M.I. Jordan, Latent Dirichlet allocation, J. Mach. Learn. Res. 3 (2003) 993–1022.

[53] J. Chang, J. Boyd-Graber, S. Gerrish, C. Wang, D. Blei, Reading tea leaves: how humans interpret topic models, in: Y. Bengio, D. Schuurmans, J. Lafferty, C.K.I. Williams, A.

Culotta (Eds.), Advances in Neural Information Processing Systems, vol. 22, MIT Press, Cambridge, MA, 2009, pp. 288–296.

[54] D.M. Blei, A.Y. Ng, M.I. Jordan, Latent Dirichlet allocation, J. Mach. Learn. Res. 3 (2003) 993–1022.

[55] T.L. Griffiths, M. Steyvers, Finding scientific topics, Proc. Natl. Acad. Sci. USA 101 (Suppl. 1) (2004) 5228–5235.

[56] Y. Song et al., Topic and keyword re-ranking for LDA-based topic modeling, in: CIKM'09, ACM, Hong Kong, China, November 2–6, 2009.

[57] V. Kakulapati et al., A re-ranking approach personalized web search results by using privacy protection, Adv. Intel. Syst. Comput. 2 (2016) 77–88, doi:10.1007/978-81-322-2752-6.

Smart sensing of medical disorder detection for physically impaired person

10

Sunil Tamhankar[a], Shefali Sonavane[a], Mayur Rathi[a], Faruk Kazi[b]

[a]*Walchand College of Engineering, Sangli, India;* [b]*Veermata Jijabai Technological Institute, Mumbai, India*

1   Introduction

The proposed system is designed to help the physically disabled person to detect the medical disorder and help accordingly with enabling the system for further aid or help.

This system uses smart wearable sensors having capability to sense parameters, convert the parametric data in desired form, and communicate with near gateway or server.

Today's common device, a smartphone, is used for enabling edge-computing functionality. GPS feature of this device is used to identify the location and RTC is used to record the timings. Message and call facility are used for immediate attention.

One of the key features of the system is the use of multiple sensors which can sense multiple parameters that are typically dependant on each other. This is used to differentiate between medical disorder and attack or malfunctioning of the sensors. This functionality improves performance of the system and reduces the false acceptance ratio (FAR).

The motivation behind the proposed system is to assist the physically imparted person in panic state with different physiological attributes. The system is validated with multivalued sensor information to infer the state of the body with age. Overall system is designed with three sensors to measure parameters, namely pulse rate, skin conductance, and respiration. Smart sensors used in this system are having capability to communicate with short-distance protocols like Bluetooth, BLE, Zigbee as IEEE 802.15.4 standard.

Gateway is the main interface between physical system and cyber system that converts short-distance protocol packet to TCP/IP packet and sends it to the server or cloud platform for storage and further processing. The proposed gateway is having multiple channels to take care of data delivery mechanism. As per the captured parametric values, data need to be polled, streamed, or necessarily delivered

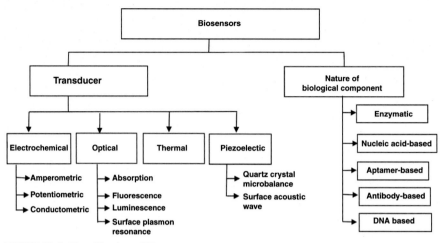

FIGURE 10.1 Classification of biosensors.

in continuous form. In the gateway, separate channels are available for desired data delivery. This multichannel multi-protocol gateway is used to reduce size and cost to improve the system performance.

This system requires large and flexible storage for captured data and high computing power for implementation of complex algorithms in decision-making solutions. The multiple attributes are logically compared to decide the state of the body in sensor network. Cloud computing platform is the best suitable environment for such kind of systems due to its availability.

Biosensors are categorized according to transducer type and the nature of biological components. These can broadly be classified as shown in Fig. 10.1.

2 Literature survey

To arrange the experimental setup and execution methodology, the domain specific work is surveyed to carry out as literature. The similar systems are studied to know the algorithms, data transmission, and data analysis in body sensor networks with quality parameters of sensor network like communication range, response time, sensitivity, etc. The communication in body sensor network is vital aspect of link transmission [1]. The energy consumption in multiparametric sensor network is discussed to improve the durability of transmission. The energy model for randomization in sensor network is specified with optimistic range of communication [2]. The preemptive modeling of behavioral analysis in health monitoring system is recommended to enhance the scope of computation model in personal healthcare system [3]. The wireless communication for bio-pressure is experimented to know the heart pulses through arteries on wristbone [4].

The heterogeneous system with sensor interoperability is discussed to recommend the low power network in health monitoring system [5]. The situation monitoring through blood flows in biomembrane gives implementable schema in body sensor network [6]. A multivalue deterministic system is endorsed to correlate the heart rate (HR), skin conductivity (SC), and facial expression for reliable and optimistic emotion detection [7,8]. The algorithmic implementation of data transmission in perspective to medical crisis and early detection of the very cause in the body for physically disabled person requires the secure platform for communication and storage [9].

Various open-source tools are available for eye tracking to detect drowsiness in mobile-based interactive systems [10,11]. The cloud-based secure and privacy preserving framework is emergent in mobile healthcare. The HR monitoring and wellness for different physical activities through various age groups with least power consumption is remarkable in health sector [12,13]. The challenges in biomedical systems are solved with implantable biosensor with secure communication protocol [14,15]. The SC for intra-body communication and challenges are defined with galvanic response in the body [16,17]. The cooperative intra-body network in body circuit is proposed for smooth data transmission and node placement for significant location-based communication [18].

The wireless medium is prone to few throughputs and optimized with cooperative routing mechanism in multi-hop environment is applicable for body sensor network. This normalizes the overhead in network and increases the packet delivery ratio significantly [19,20]. The cloud based for smart home system incorporates IOT systems to coordinate the network with gateway communication. A self-power IOT-based system is implemented to provide the medical assistance in panic situation [21]. The healthcare applications are systematically discussed to encourage IOT-based schema in patient health monitoring system [22]. Comparative study of routing mechanism in quality services of network is discussed with real-time issue of data transmission and application [23,24].

The data dependency is the major issue and is addressed with intra-routing for quality service offered by biosensor network [25]. The IEEE communication protocol 802.15.6 is utilized as a medium to assure the accuracy and reliability in link-to-link transmission [26,27]. The signal communication between the sensors is addressed with data link layer protocols for significant analysis to apply situation specific protocol in physical communications [27]. Many short-range wireless protocols are discussed in place like Bluetooth, BLE, Zigbee, 6LowPAN integrated with 802.15.4 standard [27]. The dynamic scheduling of packet for intra-sensor network is recommended for reliable communication [28]. The cardiovascular activities are monitored using the mobile device and headphones [29]. Sweat rate of the skin under various stress actions is observed to derive the activity pattern in day-to-day life using health monitoring system [30].

Table 10.1 gives the comparative review of biosensors based on its working principle, measurable useful parameters, system integration with merits and demerits [31].

Table 10.1 Comparative review of biosensors.

Sensor type	Principle	Measured property	Sensitivity	Integration
Potentiometric	Resistance change	Potential/voltage	Less sensitive	FET enzyme
Amperometric	Amount of current as a function of time by using oxidation and reduction	Current	High sensitive	Biotic recognition element, enzyme, nucleic acids, and immune sensor
Conductimetric	Proficiency of an electrolyte to manage an electrical current amongst electrodes	Electrical conductance	Relatively low	FET enzyme
Optical	Internal reflection, Principle of optical measurements like fluorescence, optical fiber, and biological recognition molecules	Surface plasmon resonance	High sensitivity	Enzyme, antibody, nucleic acids, animal cells
Piezoelectric	Sound vibrations, Quartz crystals oscillation at defined frequency	Changes in pressure, acceleration	High sensitivity	Antibody, enzymes

Table 10.2 Biosensors merits and demerits.

Merits	Demerits ·
Portable, small in size, low cost	Highly dependent on chemical properties of the sample
Availability in wide range	Temperature and humidity dependent
Less response time	Criticality in measurement
Highly sensitive and stable	Domain expertise is required to analyze the samples

Wireless sensor network for e-healthcare application is proposed in provision of medical facility from remote server with GPS location. The reliability and robustness of the system is remarkable to attend the patient irrespective of geographical position [32]. The sensitivity of wearable sensors is measured with parametric value to provide the fundamental solution with respect to the challenges in biosensor network [33].

Table 10.2 gives the merits and demerits of biosensors. The buffer capacitance of the sensor is increased to reduce the packet loss in the wearable biosensor network. To attenuate and smooth data with low pass filter, signal pre-processing is explained with various methodologies. The suitable high pass filter is introduced to record the maximum amplitude signal processing at sink node in sensor network [34].

3   System design considerations

Though sufficient literature is available describing the variety of biosensors, system needs a proper integration of instrumentation, communication, and its control. In view of this, required technical aspects are overviewed in this section.

3.1   Data collection

In Smart Body Sensor Network (SBSN) data are streamed from differentiable sensors with body parts of a user. To recollect the accurate inference, data from various sensors are interrelated to monitor the action of the user. All biosensors are interconnected in the network significantly to reflect the physiological performance. The data are measured from HR, body temperature, and sweat sensor to know the behavioral performance of the user. However, major issues are attributed around the data redundancy and replay from same sensor with respect to the time in data collection process. Collected data at gateway cannot be used directly because of various physiological factors in the sensor network.

The physical characteristics of network are supposed to greatly affect with location, motion, and environment. A network consumes more energy and time to vector valid data for processing and hence decreases the lifetime of SBSN network. By looking to these factors, data from various sensors may get secure interfere leading to further preprocessing. In SBSN, received signals are distorted due to variance in movement, gesture, location, and connectivity. These factors are to be handled crucially while experimenting the system.

The collected data are transformed to perform the computation to categorize the information statistically with variance, root mean square, time or frequency domain, wave forms, and so on. To extract the feature vector from raw data, a mathematical model can be outfitted to perform the computation. Based on sensing, location, and posture of the user, data streaming may get distorted value for the user.

In the development of SBSN, a smartphone is useful to collect data of sensors at different time intervals. This leads to another challenging issue in terms of data security. To overcome with probable threats, the multilevel data collected at gateway can be made secure by transferring it into public cloud. This helps to quarantine the fetched data or processed information against user queries. However, still the possibility of the attacks like wormhole, gray hole, or sinkhole at link interface of nodes may harm the network.

3.2   Data aggregation

Due to high density, redundant data packet gets sensed; data aggregation is a crucial aspect of body sensor network. The data collected at sink node lead to a low performance and high energy consumption that ultimately reduces the lifetime of the system.

In any sensor network, communication link is a vital medium to transfer data with transport protocol as shown in Fig. 10.2. Link communication is usually followed

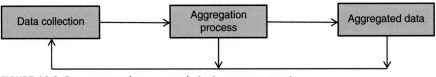

FIGURE 10.2 Data aggregation process in body sensor network.

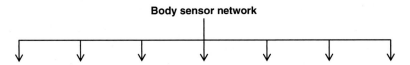

Transmission NAT Data consistency Latency Visualization Multi-tenancy Response

FIGURE 10.3 Network performance attributes.

with performance standard in order to increase the efficiency of the system. Authentic data communication can be achieved with transport layer protocol in a dedicated system. Repetitive data transmission, security, and consistency are the major factors in wireless sensor networks.

As listed in Fig. 10.3, Network Transmission Efficiency (NTE), Latency, Network Active Time (NAT), Data Consistency, Multi-tenancy, Flexibility, and Response Time (RT) are the attributes to measure the network performance to maximize the accuracy of the system. [35]

3.2.1 Transmission efficiency

It is defined as a ratio of data successfully transferred to the total energy consumed. This attribute gives the real-time usage of energy in data transmission [23].

$$\text{Transmission efficiency} = \sum_{i=1}^{n} \frac{\text{Amount of data transferred}}{\text{Energy consumed in transmission}}$$

where n is number of nodes in body sensor network.

3.2.2 Network active time

NAT can be defined as the time required to complete data aggregation for a node until energy of any one of the nodes get exhausted [23].

$$\text{NAT} = \text{Min}(\text{NAT}_i)$$

where NAT ends when first node exhausts energy.

NAT$_i$ is Network Active Time for ith node such that $i \in I$.

I is a set of nodes excluding the sink node in body sensor network.

3.2.3 Data consistency

Data consistency is the success rate of the absolute data transferred to the total amount of data sent.

$$\text{Data consistency} = \frac{\text{Data delivered successfully}}{\text{Link data transmitted}}$$

3.2.4 Latency

Latency in sensor network is the resultant delay in successive transmission of data.

$$\text{Latency} = \sum_{i=1}^{n} \text{Data reception-propogation delay}$$

3.2.5 Visualization

Data visualization is a technique to represent query data from data store in different formats [25].

3.2.6 Multi-tenancy

Multi-tenancy is an operational model referred to the independent instance of a user sharing the same storage platform [25].

3.2.7 Quick response

Quick response is the minimum delay in high computational cloud-based body network to serve the user query or to visualize the data.

In intra-network, aggregation of data at sink node is with summation, mean or weighted summation, extreme function (min, max). These techniques are to be implemented in data storage and visualization [26].

3.3 Data storage

Body sensor network receives data for various time series. Analyzing these data creates several challenges as every sensor has different sensing pattern. Data streamed from nodes cannot be handled or managed by these nodes in intra-wireless sensor network. Cloud service provides the huge storage capacity and processing platform which can be linked with body sensor network (Fig. 10.4). As data storage at gate-

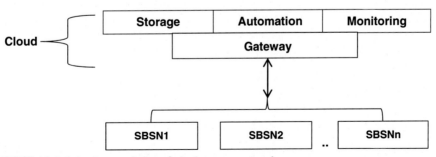

FIGURE 10.4 Data storage platform in body sensor network.

way of network is limited, cloud platform allows streaming of data to get accommodated in application perspective [26].

A system without cloud perhaps suffers from diversity due to motion, location, and change in environment. Hence, with cloud "On Go" patient can move with system without any performance barrier. This type of system is highly scalable with cost-effective solution.

Usage of cloud storage achieves following tasks.

3.3.1   Data analysis

Cloud storage accumulates the huge data stream for various time series at scalable resources for processing and analysis. The data storage can be made readily available with consistent platform compatible with the sensors in the network.

3.3.2   Scalability

Sensor network with cloud integration allows large number of nodes to forward data and accommodate with specific patterns. The storage on the cloud can be extended as per the number of nodes and users get increased.

3.3.3   Collaboration

Cloud storage can be utilized for more than one body sensor networks in sharing, irrespective of geographical location. The number of users can get access through the same server in multi-tenant environment that increases the collaborative assistance with isolation and privacy on the cloud.

3.4   Automation

The data polling and storage tasks are automated with cloud in required layout. This effort reduces the manual work to organize data on the destination server instance. The tools are auto-configured with instance on the cloud to classify the user's specific data.

3.5   Monitoring

The wearable body sensor network has to be monitored in the need of reliable medical assistance to the user. User activities are to be watched continuously irrespective of his/her location and that can be made possible with cloud platform. The multiple tenants are effectively managed with virtual environment for Platform as a Service (PAAS).

4   Proposed system architecture

Based on the discussion, a system is proposed including of major components [36] as shown in Fig. 10.5.

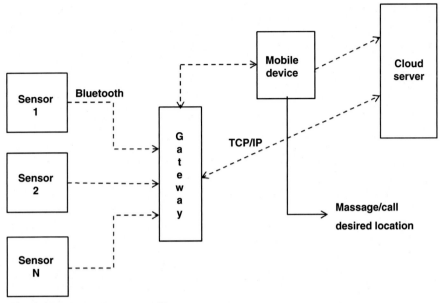

FIGURE 10.5 General system architecture.

4.1   Sensing system

Multivalue parameter needs to be measured for detection of abnormality on the non-steady state condition. Time series value for range between minimum and maximum is supposed to be collected as a real-time data. Signal pattern gets changed depending on the physiological factor. Hence, the design of a system should address the tolerance value of signals without compromising the accuracy [3].

4.2   Communication

For a given time interval, the sensor data that are placed at cloud server can get monitored and processed. Short-distance routing protocol needs to be considered to avoid the propagation delay with data compatibility at the destination server. These short-distance protocols can be from wired or wireless media of access like Modbus, Zigbee, Bluetooth, etc.

In such systems, total round-trip time and absolute data are very small; communication protocol selection should be with less control packet and more data packet on link.

The real-time data from multiparameter value system need to transit from intranet to cloud servers. Storage and processing gateways with short-distance TCP/IP are major concerns in such communication systems. Similarly, these devices should be capable to increase the packet delivery ratio as per the application requirement. Data can be continuous (live), polled, or streamed that requires versatile and multichannel gateway device to carry the task. This significantly reduces the retransmission and dissemination delay.

4.3 Skin conductivity sensor

Typically, skin resistance varies with sweat gland activity. This electrical property of skin is a vital measure to judge the specific health condition of the user.

$$Conductance = \frac{1}{Resistance}$$

The skin conductance can be measured through electrode current.

5 Attack in WSN

Wireless network is more prone to the attack due to its dependency on the communication medium. Attacks are maliciously introduced within the network to degrade the performance. Black hole, Sinkhole, Replay, and Sybil are the commonly observed attacks in the body sensor network.

As shown in Fig. 10.6, node A sends a route request to node B, node C, and node M. Node M acknowledges the route discovery to node A with high capacity. For consecutive attempt, node A follows node M for further data transmission. Unfortunately, node M is a sink node that attracts other nodes to forward the data. This reveals the data confidentiality by malicious node in the network [4].

Apart from Sinkhole attack, the other threats are also very serious in body sensor networks. As number of nodes in such networks is very limited, any loss or alterations of data are proved harmful to the health prediction and care systems. This threat can be minimized by using a secure gateway channel in the embedded system.

As an experiment, Sinkhole attack in WSN scenario is implemented in Netsim research v11.0 for analysis purpose. Table 10.3 outlines the simulation parameters used and Fig. 10.7 gives the simulation canvas.

The encircled malicious node (node 6) is in place to attract other nodes in route discovery process. The attack can be identified with number of received packets at the target node.

In Fig. 10.8, there are zero packets received at destination. This clearly indicates the interruption in the source to the destination route.

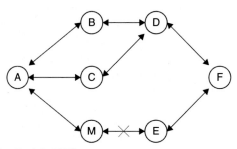

FIGURE 10.6 Sinkhole attack in WSN.

Table 10.3 Simulation parameter.

Parameter	Value
Simulation area	500 × 500 m
Simulation time	3600 s
Medium	Wireless 802.11.4
Number of nodes	06
Application	CBR
Transmission range	30 m
Routing protocol	DSR
Energy model	Generic
Massage size	36 byte

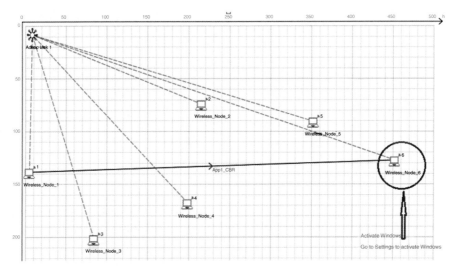

FIGURE 10.7 Sinkhole attack scenario.

Application id	Application name	Source id	Destination id	Packet generated	Packet received
1	App1_CBR	1	6	4750	0

FIGURE 10.8 Network data transfer information.

6 Proposed setup

The proposed setup is the very requirement in biosensor network to improve the life of user. The given system is suitable for average age group. Body sensors are like respiration sensor, pulse monitor sensor, and SC. The sensor network is utilized to know the state of the user consistently with time and location. Integrated system is

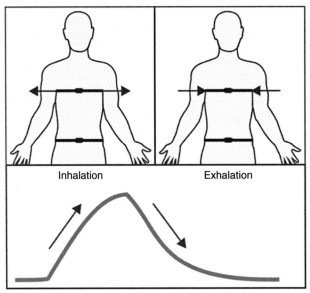

FIGURE 10.9 Inhalation and exhalation in respiration.

proposed for a blind person with sensor stick having ultrasonic sensor and wearable body sensors on the body.

6.1 Respiration sensor

Inhalation and exhalation are the functioning parameters for the judgmental record of the respiration. Fig. 10.9 shows some sample activities and related observations that can be interpreted across time intervals for measuring amplitude rise (normally mm) in the abdomen dimension.

6.1.1 Sensor principle

Based on stretching, sensor strap undergoes through rise and fall or expansion and contradiction into signal pattern from abdominal area. The signal pattern is observed for a chosen duration of time to know the respiration in differentiable situation of work done.

6.1.2 Sensor placement

The sensor strap is normally placed at abdomen or optionally on the chest. Sensor can be placed at both the places to achieve the accuracy from abdomen and chest rise or fall pattern.

6.2 Heart rate sensor

HR can be measured with sensor to know the physiological state of the body at extreme value (min/max) amplitude (Fig. 10.10).

FIGURE 10.10 Heart rate monitor (PRM).

6.2.1   Operational principle

HR sensor detects the highest amplitude for the given samples. The precision in HR can be achieved through significant number of samples for some duration of time.

6.2.2   Sensor placement

Sensor with strap should be placed against the wristbone as shown in Fig. 10.10 with body or skin. Strap of sensor should be significantly tight under heavy activity and enough loose during dormant state to measure the accurate HR [37].

6.3   Skin conductivity

The SC is a measure to define the stress or workload or activity of a person. More the conductivity, greater is the stress on the body of a person.

6.3.1   Operational principle

As skin conducts electricity, a small potential is to be applied across the palm. This helps to complete the circuit to carry the charge from one point to another point. Depending on the nervousness, person's behavior under the stress is notified to the system.

6.3.2   Sensor placement

The SC sensor can be placed at fingers (Fig. 10.11), toes, or at a palm according to the ease of the user. This can accurately be measured with the placement of circuit at palm of the hand to have maximum palm surface area for the accuracy.

6.4   Smart stick for blind

The aiding device in the form of stick is designed for blind personnel. Based on respiration sensor, pulse monitor sensor and SC of a person; signals are to be used to monitor the physiological state of the body. Fig. 10.12 demonstrates the idea. Fig. 10.13 is the practical build of a stick in the laboratory. Though, sensors are placed and showcased over the stick; can be accommodated inside the stick in a commercial product.

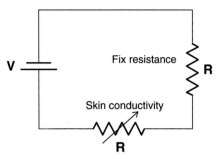

FIGURE 10.11 Skin conductivity circuit.

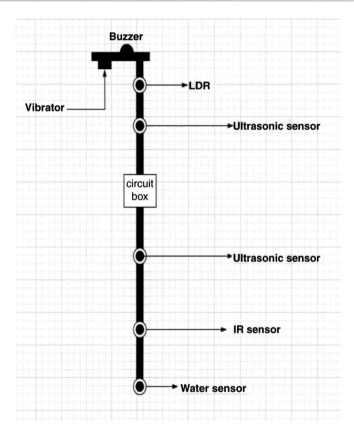

FIGURE 10.12 Smart stick prototype.

The smart stick is useful for detecting far and near obstacles in the path. The obstacle distance is expected to be automatically measured by sensors to avoid the collision of a blind in the path. This device aid is integrated with body sensor network attributed to determine the panic stage of the person as well. The values of the sensed parameters are gender and age dependent and are used for initial calibrations.

FIGURE 10.13 Smart stick for blind.

The smart stick equipped with other sensors helps the person for flawless movements. Water sensor helps to identify water on the road or slippery road conditions and alert the person while walking to change the path or act accordingly. IR and ultrasonic sensors are used to detect any obstacle in the traveling path and sensors are placed in such way that they can be able to indicate size and position of obstacle in travel path even if it is hanged at few feet above the ground level. The alarms in terms of vibrations or sound beeps are set to differentiate these conditions. The purpose of very LDR sensor is to just inform the user about the surrounding light conditions that possibly affect the sensor working. This may help the blind to take extra precautions rather than relying on the light sensors.

As shown in Table 10.4, lower and upper thresholds for sensor parameters are mentioned for the age greater than 10 years. The dependency between these sensors is linear that significantly improves the reliability to judge the state of the user [7,8].

The selection of three parameters is due to their relationship in parametric values and availability of wearable sensors that fit in the design. The heart normally beats at 60–90 times per minute and breathing rate is about one-fifth of that. The skin conductance depends upon these two parameters, because of increase in HR leads to sweating and that affects the skin conductance value. This relationship is used to identify the abnormal behavior of sensor and actual parameter change due to medical disorder.

Table 10.4 Body parameter range.

Threshold Sensor	Lower threshold	Upper threshold
Respiration rate	12 breath/min	20 breath/min
Heart rate	60 beats/min	100 beats/min
Skin conductivity	2 μs	20 μs

7 Result and analysis

The results are shown for three parameters pattern. The changes in parametric values are used to identify sensor abnormality and actual medical disorder. Fig. 10.14 shows normal signal patterns.

Fig. 10.15 shows values from a single sensor that are changing abruptly and indicating sensor abnormality.

In Fig. 10.16, parametric value change is reflected in all the three sensor values and that leads to the detection of medical disorder correctly.

These parameter values are dependent upon whether person is taking rest or walking or exercising as per the set threshold values measuring these parameters over the time and making the decisions based on the mean and deviation, calculations are used for further controls.

The algorithm is proposed for signal preprocessing and support for decision-making.

```
Algorithm
%% pre-processing
Stream input signal i = 1 to n
RRᵢ, HRᵢ, SCᵢ for time t;
Pre-process Signal
LPF (RRᵢ, HRᵢ, SCᵢ); LPF, Low Pass Filter
Down Sample
Peak Extraction
HPF (Down Sample); HPF, High Pass Filter
Classification
History Data Analysis RRᵢ, HRᵢ, SCᵢ;
Data Pattern Analysis RRᵢ, HRᵢ, SCᵢ;
%% Decision Making
If(RRi&&HRi&&SCi = =1)
Wellness State;
Else
Panic State;
```

The data collected at cloud are employed with algorithm as illustrated earlier. The received signals are operated for LPF to attenuate the pattern and smooth the

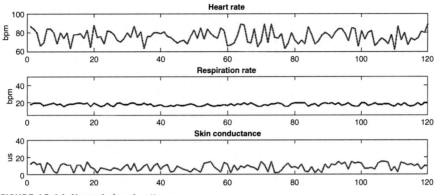

FIGURE 10.14 Normal signal patterns.

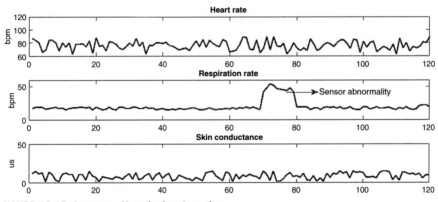

FIGURE 10.15 Sensor malfunctioning detection.

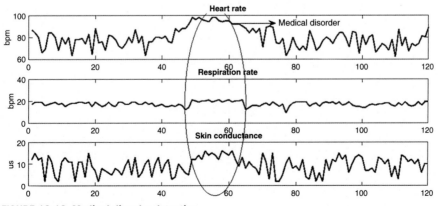

FIGURE 10.16 Medical disorder detection.

signals for data analysis. Peak value of the signal is extracted from HPF to ensure the highest maximum amplitude achieved. Data are classified with historic analysis on cloud to know the behavioral pattern of the user under certain circumstances. The data received for respiration rate(RR), heart rate(HR) and skin conductivity(SC) for a period of time are logically checked for true threshold value to know the wellness or panic state of the user.

The designed system is tested with person having middle age. The RR, HR and SC are measured for normal state and panic state. The time period is set for 10 min to record the streamed data on cloud.

7.1 Normal state

The RR, PR, and SC sensors are located as shown in Figs. 10.9–10.11, respectively, and are measured for a person. The person is under the observation with three attributes to validate the multivalued dependent statistics simultaneously.

As shown in Table 10.5, the user is under the observation in low noise location. The sensor recording during journey for a blind person is recorded and processed on cloud. The geographic location of the person is recorded with mobile GPS to reach out in emergency. The min and max value for every sensor is processed to check the dependency in the parametric value. All data values recorded are validated with each other as per the standard metric shown in Table 10.4 on cloud.

7.2 Panic state

Table 10.6 indicates the detection of the panic situation with the help of logical dependency between sensors. The sensor parameters are changed linearly with each other. The location of the user is sent to autodial number for further needy precaution.

Table 10.5 Body parameter range for normal state.

Parameter	Time slot	Record range		Activity	Location
		Min	Max		
RR	10 min	12	17	Walking	Current GPS point
HR	10 min	98	101		
SC	10 min	3	9		

Table 10.6 Body parameter range for panic state.

Parameter	Time slot	Record range		Activity	Location
		Min	Max		
RR	10 min	6	8	Walking	Current GPS point
HR	10 min	62	72		
SC	10 min	18	24		

Table 10.7   Body parameter range for sensor abnormality.

Parameter	Time slot	Record range		Activity	Location
		Min	**Max**		
RR	10 min	6	8	Walking	Current GPS point
HR	10 min	79	95		
SC	10 min	4	6		

The autodial number is activated to show the response of panic stage to the carer of the user.

7.3   Abnormal state

As shown in Table 10.7, the RR values are changing from min to max showing abnormal sensor behavior. The very condition is described in algorithm to declare the malfunctioning of the sensor and not the health problem of the user.

These are the test cases to validate the system response with high end-to-end reliability in service provision with body sensor network and storage platform.

8   Remarks

This chapter compares various biosensors. It focuses on how those sensors can be used along with wearable body sensors within a system. Use of multiple sensors clearly differentiates between actual medical disorder and corresponding sensor malfunctioning or abnormality. The data acquired need to be preprocessed and stored at cloud for its secure access. The application is devised for a blind person with wearable sensors and smart stick. The body conditions and related parameters' relationship is monitored to identify the abnormal or panic situation of the blind using available smart mobile device. This further helps to identify the location of the blind. Thus, the system may undertake immediate action ensuring timely medical assistance.

9   Future scope

In this chapter, the RR, HR, and SC are the metric observations to decide the physiological state of the physically impaired person. The proposed work is focused for blind person that can be extended to other domain of disorders with age. The medical assistance for the panic situation can be added in existing healthcare system. Currently, the system is tested for the single network and can be extended to the multiple tenants over the cloud in the distributed network.

References

[1] A.M. Nia, M. Mozaffari-Kermani, S. Sur-Kolay, A. Raghunathan, N.K. Jha, Energy-efficient long-term continuous personal health monitoring, IEEE Trans. Multi-Scale Comput. Syst. 1 (2) (2015) 85–98.

[2] N. Dam, A. Ricketts, B. Catlett, J. Henriques, Wearable sensors for analysing personal exposure to air pollution, in: Systems and Information Engineering Design Symposium (SIEDS), IEEE, Charlottesville, VA, USA, 2017, pp. 1–4.

[3] M. Pavel, H.B. Jimison, I. Korhonen, C.M. Gordon, N. Saranummi, Behavioral informatics and computational modeling in support of proactive health management and care, IEEE Trans. Biomed. Eng. 62 (12) (2015) 2763–2775.

[4] Y. Peng, T. Wang, W. Jiang, X. Liu, X. Wen, G. Wang, Modeling and optimization of inductively coupled wireless bio-pressure sensor system using the design of experiments method, IEEE Trans. Compo. Packaging Manufac. Technol. 8 (1) (2018) 75–79.

[5] H. Fotouhi, A. Causevic, M. Vahabi, M. Bjrkman, Interoperability in heterogeneous low-power wireless networks for health monitoring systems, in: IEEE International Conference on Communications Workshops (ICC), IEEE, 2016, pp. 393–398.

[6] C. Steeves, Y. Young, Z. Liu, A. Bapat, K. Bhalerao, A. Soboyejo, W. Soboyejo, Membrane thickness design of implantable bio-MEMS sensors for the in-situ monitoring of blood flow, J. Mater. Sci. Mater. Med. 18 (1) (2007) 25–37.

[7] J. Lei, Identifying correlation between facial expression and heart rate and skin conductance with iMotions biometric platform, J. Emerg. Forensic Sci. Res. 2 (2) (2017) 53–83.

[8] S. Misra, S. Sarkar, Priority-based time-slot allocation in wireless body area networks during medical emergency situations: an evolutionary game-theoretic perspective, IEEE J. Biomed. Health Inform. 19 (2) (2015) 541–548.

[9] T. Wyss, The comfort acceptability and accuracy of energy expenditure estimation from wearable ambulatory physical activity monitoring systems in soldiers, J. Sci. Med. Sport 20 (2017) S133–S134.

[10] H. Suryotrisongko, F. Samopa, Evaluating OpenBCI spiderclaw V1 headwear's electrodes placements for brain-computer interface (BCI) motor imagery application, Procedia Comput. Sci. 72 (2015) 398–405.

[11] M. Kassner et al., Pupil: an open source platform for pervasive eye tracking and mobile gaze-based interaction, in: Proceedings of the 2014 ACM International Joint Conference on Pervasive and Ubiquitous Computing Adjunct Publication, ACM, 2014, pp. 1151–1160.

[12] R.X. Lu, X.D. Lin, X.M. (Sherman) Shen, SPOC: a secure and privacy-preserving opportunistic computing framework for mobile-healthcare emergency, IEEE Trans. Parall. Distr. 24 (3) (2013) 614–624.

[13] D. Bortolotti, M. Mangia, A. Bartolini, R. Rovatti, G. Setti, L. Benini, An ultra-low power dual-mode ECG monitor for healthcare and wellness, in: Proceedings of the Design, Automation & Test in Europe Conference & Exhibition (DATE), IEEE, Grenoble, France, 2015, pp. 1611–1616.

[14] G.E. Santagati, T. Melodia, Sonar inside your body: prototyping ultrasonic intra-body sensor networks, in: Proceedings of the IEEE Conference on Computer Communications, IEEE, Toronto, ON, Canada, 2014, pp. 2679–2687.

[15] A. Kiourti, K.S. Nikita, A review of implantable patch antennas for biomedical telemetry: challenges and solutions, IEEE Antennas Propag. Mag. 54 (2012) 210–228.

[16] M. Callejon, D. Naranjo, J. Reina-Tosina, L. Roa, Distributed circuit modeling of galvanic and capacitive coupling for intrabody communication, IEEE Trans. Biomed. Eng. 59 (11) (2012) 3263–3269.

[17] B. Kibret, M. Seyedi, D.T.H. Lai, M. Faulkner, Investigation of galvanic-coupled intrabody communication using the human body circuit model, IEEE J. Biomed. Health Inform. 18 (4) (2014) 1196–1206.

[18] M.T.I. ul Huque, K.S. Munasinghe, A. Jamalipour, Body node coordinator placement algorithms for wireless body area networks, IEEE Internet Things J. 2 (2015) 94–102.

[19] Q. Nadeem, N. Javaid, S.N. Mohammad, M.Y. Khan, SIMPLE: stable increased-throughput multi-hop protocol for link efficiency in wireless body area networks, in: Broadband and Wireless Computing Communication and Applications (BWCCA) IEEE Conference, IEEE, Compiegne, France, 2013, pp. 221–226.

[20] F. Wu, J. Redout, M.R. Yuce, We-safe: a self-powered wearable IoT sensor network for safety applications based on LoRa, IEEE Access 6 (2018) 40846–40853.

[21] D. Antolin, N. Medrano, B. Calvo, F. Pérez, A wearable wireless sensor network for indoor smart environment monitoring in safety applications, Sensors MDPI 17 (2) (2017) 365.

[22] T. Wu, F. Wu, J.-M. Redoute, M.R. Yuce, An autonomous wireless body area network implementation towards IoT connected healthcare applications, IEEE Access 5 (2017) 11413–11422.

[23] M. Anand Kumar, C. Vidya Raj, A comprehensive survey of QoS-aware routing protocols in wireless body area networks, Int. J. Adv. Res. Comput. Commun. Eng. 6 (1) (2017) 166–173.

[24] M.M. Monowar, M.M. Hassan, F. Bajaber, M.A. Hamid, A. Alamri, Thermal-aware multiconstrained intrabody QoS routing for wireless body area networks, Int. J. Distrib. Sens. Netw. 10 (2014) 676312:1–:1676312.

[25] S. Gao, G. Ver Steeg, A. Galstyan, Efficient estimation of mutual information for strongly dependent variables, Artif. Intell. Stat. 38 (2015) 277–286.

[26] K.S. Deepak, A.V. Babu, Improving reliability of emergency data frame transmission in IEEE 802.15.6 wireless body area networks, IEEE Systems Journal (Early Access), 2017, 1–12.

[27] A. Horrich, L. Issaoui, K. Sethom, Improved MAC access under IEEE 802.15.6 WBAN standard, in: International Conference on Ubiquitous and Future Networks (ICUFN), IEEE, Milan, Italy, 2017.

[28] I. Romdhani, N. Salayma, A. Al-Dubai, Y. Nasser, New dynamic reliable and energy efficient scheduling for wireless body area networks (WBAN), in: IEEE International Conference on Communications (ICC), IEEE, Paris, France, 2017.

[29] M.Z. Poh, Cardiovascular monitoring using earphones and a mobile device, IEEE Perv. Comput. 11 (4) (2012) 18–26.

[30] S. Patel, Monitoring motor fluctuations in patients with Parkinson's disease using wearable sensors, IEEE Trans. Inform. Technol. Biomed. 13 (6) (2009) 864–873.

[31] C. Poon, Y.-T. Zhang, S.-D. Bao, A novel biometrics method to secure wireless body area sensor networks for telemedicine and m-health, IEEE Commun. Mag. 44 (4) (2006) 73–81.

[32] P. Castillejo, Integration of wearable devices in a wireless sensor network for an e-health application, IEEE Wireless Commun. 20 (4) (2013) 38–49.

[33] J. Cheng, et al. Designing sensitive wearable capacitive sensors for activity recognition, IEEE Sensors J. 13 (10) (2013) 3935–3947.

[34] H.N. Teodorescu, M. Hagan, Signal processing for a wearable device for activity monitoring, in: Proceedings of the International Conference on E-Health and Bioengineering EHB, IEEE, Sinaia, Romania, 2017.

[35] M.A. Hannan, S. Mutashar, S.A. Samad, A. Hussain, Energy harvesting for the implantable biomedical devices: issues and challenges, Biomed. Eng. Online 13 (2014) 79.

[36] Sunil Tamhankar, Faruk Kazi, Sachin Patil. Design of SMART (Secure, Multichannel, Adaptive, Real Time, Tiny) Gateway for Cyber Physical System. Int. J. Comput., **3** (2018) 20–26.

[37] P. Salvo, et al. A wearable sensor for measuring sweat rate, IEEE Sensors J. 10 (10) (2010) 1557–1558.

Speech-based automation system for the patient in orthopedic trauma ward

11

Dharm Singh Jat[a], Anton S. Limbo[a], Charu Singh[b]

[a]Namibia University of Science and Technology, Windhoek, Namibia; [b]Sat-Com (PTY) Ltd., Windhoek, Namibia

1 Introduction

Speech recognition (SR) is the use of voice inputs into a computing device. The device converts the speech signal to a format that computers can process [1]. This convention normally requires the analog signal to be represented into a digital format by processing features of the speech to determine the most likely words or sentences contained in the spoken speech.

In recent years, SR technologies have become more popular, and as a result SR technologies are changing the way we interact with computing devices. SR is now a popular method of input to replace input methods like typing and clicking [1]. Today, one can find SR technology embedded many sectors such as in the automotive industry enabling car drivers to control certain aspects of the car using voice commands without having to leave the steering wheel. Perhaps the most common application of SR is in modern mobile phones, where most come with an SR engine that enables the user to be able to control the device using voice commands; this includes features like voice searching and dialing contacts, looking up directions using maps, among other things. Other applications of SR include voice-enabled virtual assistants with prominent ones like the Amazon Alexa, Google Home Assistant, and Apple Home Pod [2].

The advancement in SR has also been complemented with the increase in Internet of Things (IoT) applications. IoT technology has enabled us to be able to connect physical devices to the internet, enabling these devices to generate data and transmit these data to other devices [3]. This has also led to an increase in sensors that can be used to measure certain parameters without the presence of human being, and then use these measurements to automate certain processes.

Despite these advancements, there is still little application of these technologies in healthcare. This chapter discusses an automation system based on spoken commands, to be placed in an orthopedic ward. During hospitalization, orthopedic patients require appropriate intervention and careful monitoring by nursing staff before and after undergoing surgery [4]. According to the Global Health Observatory data repository [5], the number of nursing staff per 10,000 population is as low as 2 in developing countries. This indicates that current nursing staff cannot fully monitor and attend to patients effectively even those that lack the mobility and need constant monitoring, like orthopedic patients.

Although hospitals increasing implementing monitoring strategies to assist patients in both automating functions like calling a nurse by pressing a button and adjusting bed elevation from a remote, these features cannot be utilized by orthopedic patients with lack of mobility. There is need to determine other methods in which patients with limited mobility can automate certain tasks while in hospital, one such approach is to enable patients to control devices in a ward using voice commands. This chapter presents a system that employs SR to enable patients in an orthopedic ward to automate tasks as well as provide a timely monitoring system for nurses to monitor multiple patients' vital signs. The use of an automation system will significantly reduce the reliance of patients on nurses and other caregivers while enabling nurses to monitor patients' vital signs in real time.

2 Related work

This section presents some work that has been done in the area of SR in healthcare, as well as some automation techniques using sensors.

Hossain [6] designed a system that used speech and video as input to determine the state of a patient. The system captures the speech and facial video of a patient and transfers the data to a cloud server with the patient's identification. The cloud server would then process these two parameters separately to extract features that can help to determine the state of the patient and give a score of these parameters. The scores then combined to output the final decision regarding the state of the patient and relayed to a healthcare staff monitoring the patient. To assess the accuracy of the system, 100 people were recruited to mimic a patient's states of normal, pain, and tensed. The experimental results show that the proposed system can achieve an average of 98.2% recognition accuracy.

Work done by Doukas and Maglogiannis [7] presents a prototype implementation of IoT and cloud technologies which employs wearable and mobile sensors to acquire patient data and transmit these data to a cloud platform. Sensors collect biosignals from the user such as heart rate, electrocardiogram, oxygen saturation and temperature, motion data through accelerometers and contextual data like location, ambient temperature, activity status. The data are then transmitted to a mobile phone or cloud platform where appropriate interfaces enable the data dissemination to external applications like medical record systems, emergency detection platforms, real-time monitoring, and management systems.

Muhammad [8] proposed a cloud framework for speech in healthcare. In the framework, a patient seeking medical assistance can transmit this request using speech to a cloud server. The speech commands are managed by a cloud server which also enables any medical doctor with authentication to the server to receive and attend to the request. The framework proposes a new feature extraction technique named interlaced derivative pattern (IDP) in automatic speech recognition (ASR) system to be deployed into the cloud server. The IDP exploits the relative Mel-filter bank coefficients along with different neighborhood directions from the speech signal. Experimental results showed that the proposed IDP-based ASR system performs reasonably well even when the speech is transmitted via smartphones.

Catarinucci et al. [9] propose a smart hospital system (SHS) which relies and compliments technologies such as radio frequency identification (RFID), wireless sensor network (WSN), and smart mobile technologies to be able to collect data in real time of both environmental conditions and patients' physiological parameters via an ultra-low-power hybrid sensing network (HSN) composed of low-power wireless personal area network (6LoWPAN) nodes integrating ultra high frequency (UHF) RFID functionalities. Sensed data are transmitted to a control center where a monitoring application makes it accessible to local and remote users via web service.

Study done by Yeh [10] shows the introduction of a secure IoT-based healthcare system which utilizes body sensor network implemented during a Raspberry Pi. The system employs crypto-primitives to form two communication mechanisms that ensure transmission confidentiality while providing entity authentication among smart devices. The study discusses security requirements for IoT-based systems with the overall objective of securing sensitive healthcare data being monitored on the patient when these data are transit.

Work done by Wu et al. [11] proposed a node with wearable sensors capable solar energy harvesting and equipped with low energy Bluetooth transmission forming an autonomous wireless body area network (WBAN). In this system, multiple sensors are placed on different parts of the body to measure parameters such as heartbeat, body temperature, falls, etc. The readings from the sensors are displayed on a mobile application and which also serves as alerting mechanisms for falls. Experimental results showed that the system can achieve 24 hours of operation if the subject stays outside for short period of time to enable the system to harvest solar energy.

The work by Mayer [12] describes an IoT architecture that can assist people with mobility restrictions to control and monitor appliances and sensors in a home. The IoT architecture employs low cost, off-shelve hardware components which are geared toward people who cannot afford a full-time helper in their homes. The architecture uses low bandwidth and only makes internet connection when necessary to further reduce the operational cost of the system. The uses of a single board computer is an IoT gateway to communicate using various protocols with appliances and sensors in a home and transmit these data to a remote server to enable users to interact with other people as well as to get information like forecast updates by using only the voice. The architecture accuracy was close to 100% as it employed noise suppression methods to address false positives and negatives and disturbances.

The study presented an implementation of an automated medical scribe assistant, capable of listening to encounters and procedures between a patient and physician and chart these encounters in real time [13]. The system uses multiple speeches and language technologies including speaker diarization, medical speech recognition, knowledge extraction, and natural language to mine for appropriate information used to produce a formatted section of the report from the conversation. Speaker diarization is used to distinguish who spoke when during the conversation.

Previous work done in fields related to this work includes the use of IoT, SR, and cloud technologies to improve healthcare systems. However, previous work done does not address the reliance of hospitalized patients on nursing staff and caregivers by not leveraging on these modern technologies to reduce this reliance. This chapter discusses a system that reduces reliance on nursing staff by patients in an orthopedic ward. The system complements and utilizes existing SR technologies with IoT devices and sensors to enable patients to automate certain processes by using voice commands, while at the same time giving a platform for nursing staff to monitor patients in a ward.

To increase the efficiency of this approach, the system relies less on internet connectivity and processes most data on the single board computers interconnected with wireless local area network (WLAN).

3   System design

This section provides a conceptual design for the edge computing devices which is capable of performing SR on a single board computer to assist patients in an orthopedic room or ward to automate a certain process by only using patients' voice without the use of cloud servers. Fig. 11.1 shows the conceptual design of the system.

In the conceptual design, an orthopedic ward is equipped with the edge computing device which in this case is a Raspberry Pi 3 single board computer. This device is responsible for performing offline speech recognition to assist the patient in automating some of the tasks in the ward. To achieve this, the Raspberry Pi is equipped with a microphone to capture voice commands in the form of speech signal from the patient. The received speech signal is processed for feature extraction before being sent to the speech decoder to determine the content of the speech signal received from the patient. The transcribed text is forwarded to the automation module which employs infrared technology to control appliances in the ward. At the same time, the Raspberry Pi transmits data are being monitored to another Raspberry Pi at the nurse monitoring station.

The Raspberry Pis are interconnected using a wireless access point to form a WLAN. The WLAN ensures that the computing device is able to feed real-time data about the patients to the nurse monitoring station. Multiple Raspberry Pis can be placed in different wards and will at regular intervals transmit data from monitoring sensors to the nurse station. These data can enable nurses to monitor multiple patients in different wards. Some information that can be useful to be displayed includes a temperature in the room, live image feed of the patient, lighting in the room, patient body temperature, heartbeat rate, etc.

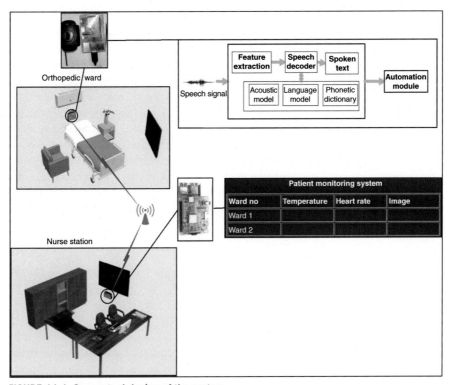

FIGURE 11.1 Conceptual design of the system.

4   System implementation

The conceptual design is implemented on two Raspberry Pis devices, where one device represents a computing node that will be placed in a ward and the second device represents the monitoring device to be placed at the nurse station. The device in the ward consists of two modules. The first module is responsible for capturing the speech from the patient and determining the action to be taken based on the spoken commands, while the second module is for initiating the automation process.

4.1   Speech recognition overview

As mentioned earlier, speech involves the conversion of spoken words into a format that can be interpreted by a computer. In this case, the system converts speech into text format.

Speech recognition systems can be classified into four types based on how they convert the speech into text. According to Renals and Shimodaira [14], SR systems can be classified into speaker-dependent systems which are trained to recognize a single speaker and speaker-independent systems which do not need the training to adapt to speakers speaking style.

Another classification of SR systems is between continuous SR systems which can recognize spoken speech as the user speaks and another type the user has to pause between words to enable an SR system to recognize isolated words.

Despite these types of SR systems, they all work in a similar manner by employing three models to perform speech recognition [15]. These include an acoustic model, which is representation of different raw audio input signals and a mapping of these signals to phonemes. Another model employed is a language model which is statistical information about which words should follow each other in a sentence. A phonetic dictionary maps the relationship between vocabulary words and phonemes.

The overall goal of an SR system is to find the maximum likely word sequence from an audio signal [14]. To achieve this, SR systems rely on the fundamental equation of statistical speech recognition, which states that if X is the sequence of acoustic feature vectors (observations) and W denotes a word sequence, the most likely word sequence is given by:

$$W^* = \arg\max_w P(W/X) \tag{1}$$

The accuracy of an SR system is measured using the word error rate (WER) which is represented by the following equation [14]:

$$WER = \frac{S + D + I}{N} \times 100 \tag{2}$$

In the equation, S represents the number of substitutions that an SR system makes by substituting a word in the transcribed text instead of the spoken word. D represents the number of deletions the system makes in the transcribed text. And I represents the number of words the SR system inserts into the transcribed text instead of what the user had said. N represents the total number of words spoken.

SR system employs two main types of algorithms to achieve speech recognition which are Hidden Markov Model (HMM) and Artificial Neural Networks (ANN). Statistical model to estimate the probability of a set of observations based on the sequence of hidden state transitions [16]. An HMM is a stochastic signal model and the simplest variant of a dynamic Bayesian network. The SR system employed in this system used HMM to perform speech recognition.

4.2 System speech recognition

To perform speech recognition, we employ open-source toolkits from CMUSphinx. CMUSphinx is developed by Carnegie Mellon University to be used for research in speech recognition [17]. The CMUSphinx project also develops PocketSphinx, which is a portable, lightweight speech recognition engine that can be implemented on devices with minimal processing power such as ARM-based processors, which includes a range of devices like a Raspberry Pi and smartphones [18].

As mentioned earlier, an SR system uses an acoustic model, language model, and phonetic dictionary to be able to convert spoken words to their text equivalence. In the case of this system, a language and phonetic dictionary was generated with

Table 11.1   Phonetic dictionary used by the system.

Vocabulary word	Phonetic representation
BED	B EH D
CALL	K AO L
CHANNEL	CH AE N AH L
COOL	K UW L
DOWN	D AW N
HEAD	HH EH D
LEGS	L EH G Z
LIGHT	L AY T
NURSE	N ER S
OFF	AO F
ON	AA N
ON(2)	AO N
TEMPERATURE	T EH M P R AH CH ER
TEMPERATURE(2)	T EH M P ER AH CH ER
TV	T IY V IY
TV(2)	T EH L AH V IH ZH AH N
UP	AH P
VOLUME	V AA L Y UW M
WARM	W AO R M

limited words that correspond to processes to be automated. Testing results have shown that using a custom phonetic dictionary and language model improves the accuracy of PocketSphinx as it reduces the amount of data to search through to determine the text equivalence of the phonemes spoken. Table 11.1 shows the phonetic dictionary used by the system for automation.

The patient has to say a sentence with the appliance they want to operate as well as the state to which they want to change the device to. From the phonetic dictionary, a patient can perform automation tasks such as switching on the TV in the ward by saying *TV On*. Most television sets have the channel up/down buttons to allow the users to scroll through various channels available; this can be done by the patient saying *TV Channel Down* or *TV Channel Up*. The volume of the appliance can also be operated in a similar manner by the patient saying *TV Volume Up* or *TV Volume Down* to increase and decrease the volume, respectively. To reduce the volume or mute the appliance the patient can say *TV Mute*.

Wards with air-conditioning units can also be operated using commands such as *Temperature Warm* or *Temperature Cool* to increase and decrease temperature in the room. Lighting in the room can be operated in a similar manner by simply saying *Light On* and *Light Off*.

To call a nursing staff to the ward, a patient can say *Call Nurse*; this voice command will initiate a connection with the module at the nurse station monitoring patients in wards and enable the Raspberry Pi at the nurse station to sound a buzzer and indicate which ward the call came from.

In cases where patient beds come with an infrared remote, the system can be integrated to enable patients to control aspects of the bed using voice commands. These aspects can be to lower and raise the height of the bed to make it easy for the patient to come on/off the bed. The patient can also voice commands to raise the head/legs section of the bed to increase their comfort. These are all features that can be very beneficial to patients in an orthopedic ward with reduced ability to use their limbs, as it will enable them to control features with the voice and reduce the reliance on caregivers and nurses.

4.3 Hardware requirement

The system employs a Raspberry Pi 3 Model B single board computer as the main computing device to perform the speech recognition and automation as well as monitoring for the nurse station. A Raspberry Pi 3 is a low-cost, credit-card-sized computer, equipped with a quad-core Cortex-A53 processor, Bluetooth, wireless LAN, as well as Ethernet connectivity. The Raspberry Pi is also equipped with General Purpose Input and Output (GPIO) pins, which enable interconnectivity of sensors to the Raspberry Pi [19].

In context of the system, the Raspberry Pi employs a USB webcam with an integrated microphone to capture speech signal and images. To be able to control appliances, the system employs infrared (IR). For this purpose, an IR transmitter and receiver are required to enable the system to be programmed to transmit specific IR signal which corresponds to the voice commands. The infrared transmitter and receiver are connected via the GPIO of the Raspberry Pi.

4.4 System automation

The system was implemented using Python programming language. The webcam captures voice commands as well as periodically takes an image of the ward; however, other sensors can easily be integrated into the system to be included at the monitoring station. Fig. 11.2 shows a flowchart on how the system works.

The following steps show the pseudo code of the system corresponding to the flowchart in Fig. 11.2.

1. Start system
2. Initialize microphone
3. While system is running:
   a. Read data from sensors
   b. Transmit data to be displayed on monitoring station
   c. Display data on monitoring screen
4. Start capturing voice commands
5. Convert voice commands into text

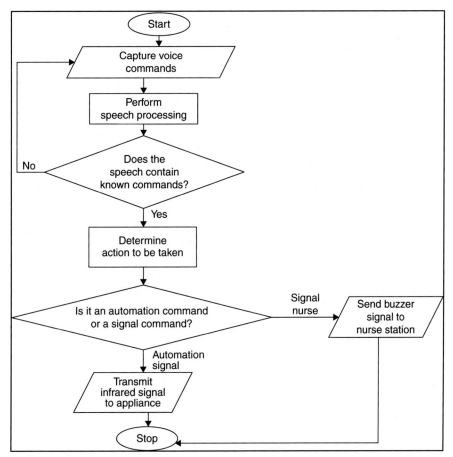

FIGURE 11.2 Flowchart of the system.

---

6. Determine the commands in the text
7. Match the spoken commands with known commands
8. Determine what action to take based on the command
9. If command contains *Call Nurse*
    a. Send signal to ring the buzzer at the nurse station
    b. Sound buzzer at monitoring station
    Else command contains an infrared-based automation.

    a. Transmit infrared signal representing the received command

10. Restart from step 2

To enable the system to integrate with existing appliances, the system uses infrared (IR) to control appliances. The system employs Linux Infrared Remote Control

(LIRC). LIRC enables the Raspberry Pi to be able to receive and transmit IR signal. The Raspberry Pi is connected to an IR transmitter and receiver via the GPIO pins.

The IR receiver is used to capture the encoding signal from an IR remote. This signal is matched to a specific voice command representing the action of the button pressed on the remote. This applies to basic remote control usage like switching a device on/off, channel up or down, and so on. To automate, when a voice command is received, the system matches the converted text and determines which infrared signal to transmit equivalent to that of a button pressed on a remote control.

In order to ring a buzzer to inform a nursing staff that the patient needs assistance, the monitoring Raspberry Pi sends a signal to sound a buzzer with ward number to another Raspberry Pi responsible for displaying monitoring information at nurse station. This communication is done using the WLAN.

In the background, the Raspberry Pis in wards read data from temperature sensor and capture an image of the patient and transmit these readings to the monitoring Raspberry Pi at the nurse station using the WLAN. The monitoring Raspberry Pi then displays these readings on the screen being observed by nurses at the station. This will give the nurses an idea of the condition of the patients in wards without having to physically do routine checks on the patients. With the advancement of sensor technologies, it is possible to measure other aspects of patients like heartbeat rate.

4.5 System monitoring

For monitoring the system uses a Raspberry Pi situated at the nurse station. This Raspberry Pi is connected to a display for the nurses to see real-time information about the patients in the wards. The information is displayed using a web browser, which also allows the nurses to be able to see this information on any web-enabled devices like their smartphone connected to the WLAN.

SSH and public-key authentication is used by the Raspberry Pis in the wards to transmit data to the monitoring Raspberry Pi. The use of SSH and public keys for authentication enables the Raspberry Pis in the wards to transfer data securely with the monitoring Raspberry Pi. The sequence of this would be the following:

1. Capture an image at every 30-minute interval
2. Read data from the sensors connected to the Raspberry Pi
3. Initiate an SSH connection to the monitoring Raspberry Pi using SSH password-less login
4. Transfer data to monitoring Raspberry Pi
5. Monitoring Raspberry Pi inserts data web server
6. Restart web server
7. Reload web browser on the monitoring Raspberry Pi

The aforementioned are only for the transfer and displaying of the data onto the monitoring Raspberry Pi, when a buzzer is initiated by the patient; this is sent immediately to the monitoring Raspberry Pi to inform the nurses that a patient might be in distress and needs their physical assistance.

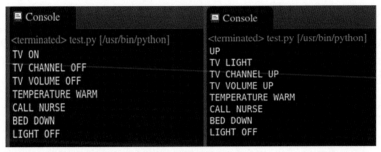

FIGURE 11.3 Results of transcribed text for Speakers A and B.

4.6   Results

To evaluate the system, the speech recognition part of the system was evaluated on two speakers who spoke various voice commands and the output of the system was recorded to measure the WER.

Fig. 11.3 shows the output of the transcribed text by the two speakers. In this case, the speakers spoke seven phrases in the same order to determine the consistence of the recognition component of the system when being used by different speakers.

Table 11.2 shows the words that both speakers spoke and the words that system transcribed to text. The system was then measured using the WER formula, measuring the amount of words the system interpreted wrongly divided by the total number words spoken.

$$\text{WER} = \frac{3}{32} \times 100 = 9.38\% \tag{3}$$

This WER indicates 90% accuracy by the system. The WER can be reduced by applying more rules in the language module to increase the accuracy of speech recognition in the system. For example, when Speaker A gave the command *TV Channel Up*, the semantics should not allow *Off* as the last word because there is *Channel*

Table 11.2   Test results for speech recognition for two speakers.

| | Transcribed text | |
Spoken phrase	Speaker A	Speaker B
TV on	TV on	TV light
TV channel up	TV channel off	TV channel up
TV volume up	TV volume off	TV volume up
Temperature warm	Temperature warm	Temperature warm
Call nurse	Call nurse	Call nurse
Bed down	Bed down	Bed down
Light off	Light off	Light off

in the middle which can only be followed by either *UP* or *Down.* This is observed as well in the other phrases where the system substitutes a wrong word in the transcribed text, for example, in Speaker B's first phrase. The system substitutes *On* with *Light* but according to semantics *TV* can only be followed with either *On* or *Off.* So this is definitely something to improve in the language model.

5 Conclusion

This chapter discussed the use of speech recognition to enable patients in an orthopedic trauma ward to be able to automate processes by using voice commands. The system is implemented on using Raspberry Pi single board computer. The system uses PocketSphinx to perform offline speech recognition of spoken voice commands to text. The system further employs LIRC to operate infrared-enabled devices in a ward. This includes television set and temperature control unit like an air-condition unit. This is done by matching the converted text with a corresponding infrared signal. The results showed that improved accuracy in speech recognition could be achieved by reducing the number of words in a phonetic dictionary and language model to only words that are needed for automation.

6 Future work

For this study we employed limited sensors into the system, but with the growing advancement in sensor technologies, there is a possibility to integrate multiple sensors onto a one single board computer to measure other aspects of a patient as well as the parameters in the environment in which the patient is being housed.

Sensors that have be added include heart rate sensor that can be connected to the patients and transmit the rate in real time to the monitoring station which can give nurses and doctors to monitor the heart condition of the patient and possibly study the patterns of this rate.

Body temperature sensor can also be added to the patients to give nursing staff a more detailed condition of the patient. The Raspberry Pi can then be configured to sound the buzzer when the body temperature of the patient rises or reduces by a certain threshold. This will give the nurse a more timely response to the ward to investigate the cause of the high or lower temperature of the patient.

Light intensity sensor can also be added to the system to measure the amount of light in the room, which can then be used to automate the switching on and off of the lighting in the ward.

References

[1] N. Unuth, What is speech recognition? Available from: https://www.lifewire.com/what-is-speech-recognition-3426721, 2019.

[2] U. Pisipati, Automatic speech recognition, virtual assistants and the rise of conversational era. Available from: https://medium.com/@udaynag/automatic-speech-recognition-virtual-assistants-and-the-rise-of-conversational-era-1493a224d4eb, 2017.

[3] Vandana, How speech-to-text/voice recognition is making an impact on IoT development. Available from: https://internetofthingswiki.com/how-speech-to-text-voice-recognition-is-making-an-impact-on-iot-development/1269/, 2018.

[4] I.C.O. Vital, L.E. Cameron, T.R. Cunha, C.I. Santos, Information as an instrument of care to patients undergoing orthopedic surgery, Cogitare Enferm. 23 (1) (2018) e51192.

[5] World Health Organization, Global Health Observatory data repository, nursing and midwifery personnel. Available from: http://apps.who.int/gho/data/node.main.HWFGRP_0040?lang=en, 2019.

[6] M.S. Hossain, Patient state recognition system for healthcare using speech and facial expressions, J. Med. Syst. 40 (2016) 272. Available from: https://doi.org/10.1007/s10916-016-0627-x.

[7] C. Doukas, I. Maglogiannis, Bringing IoT and Cloud Computing towards Pervasive Healthcare. Paper Presented at Sixth International Conference on Innovative Mobile and Internet Services in Ubiquitous Computing, Palermo, Italy, 2012. DOI: 10.1109/IMIS.2012.26.

[8] G. Muhammad, Automatic speech recognition using interlaced derivative pattern for cloud based healthcare system, Cluster Comput. 18 (2015) 795. Available from: https://doi.org/10.1007/s10586-015-0439-7.

[9] L. Catarinucci, D. Donno, L. Mainetti, L. Palano, L. Patrono, M.L. Stefanizzi, L. Tarricone, An IoT-aware architecture for smart healthcare systems, IEEE Internet Things J. 2 (6) (2015) 515–526, doi: 10.1109/JIOT.2015.2417684.

[10] K. Yeh, A secure IoT-based healthcare system with body sensor networks, IEEE Access 4 (2016) 10288–10299, doi: 10.1109/ACCESS.2016.2638038.

[11] T. Wu, F. Wu, J. Redouté, R.M. Yuce, An autonomous wireless body area network implementation towards IoT connected healthcare applications, IEEE Access 5 (2017) 11413–11422, doi: 10.1109/ACCESS.2017.2716344.

[12] J. Mayer, IoT architecture for home automation by speech control aimed to assist people with mobility restrictions, in: Proceedings of International Conference on Internet Computing and Internet of Things, CSREA Press, Las Vegas, USA, 2017.

[13] G.P. Finley, E. Edwards, A. Robinson, N. Sadoughi, J. Fone, M. Miller, D. Suendermann-Oeft, M. Brenndoerfer, N. Axtmann, An automated assistant for medical scribes, in: Proceedings of Interspeech 2018, Hyderabad, India, 2018.

[14] S. Renals, H. Shimodaira, Introduction to speech recognition [PDF file]. Available from: https://www.inf.ed.ac.uk/teaching/courses/asr/2015-16/asr01-intro.pdf, 2014.

[15] C. Woodford, Speech recognition software. Available from: https://www.explainthatstuff.com/voicerecognition.html, 2019.

[16] E. Kang, Hidden Markov Model. Available from: https://medium.com/@kangeugine/hidden-markov-model-7681c22f5b9, 2017.

[17] CMUSphinx, Overview of the CMUSphinx toolkit. Available from https://cmusphinx.github.io/wiki/tutorialoverview/.

[18] CMUSphinx, Building an application with PocketSphinx. Available from: https://cmusphinx.github.io/wiki/tutorialpocketsphinx/.

[19] Raspberry Pi Foundation, Raspberry Pi 3 Model B + . Available from: https://www.raspberrypi.org/products/raspberry-pi-3-model-b-plus/.

Index

215